Heba S. Hamed
Afaf Abo Nour
Adel Shaheen

Toxicity of Silver nanoparticles on Catfish: Effect of Nigella sativa

Heba S. Hamed
Afaf Abo Nour
Adel Shaheen

Toxicity of Silver nanoparticles on Catfish: Effect of Nigella sativa

Noor Publishing

Imprint

Cover image: www.ingimage.com

Publisher:
Noor Publishing
is a trademark of
Dodo Books Indian Ocean Ltd. and OmniScriptum S.R.L publishing group

120 High Road, East Finchley, London, N2 9ED, United Kingdom
Str. Armeneasca 28/1, office 1, Chisinau MD-2012, Republic of Moldova, Europe
Managing Directors: Ieva Konstantinova, Victoria Ursu
info@omniscriptum.com

Printed at: see last page
ISBN: 978-620-9-26438-2

DEDICATION

To my Kind

Parents

To my

Faithful Brother

Kyrillos Gamal

Courses

Courses studied by the candidate in partial fulfillment of the requirements for the degree of M.Sc.

1-Physiology

2- Histology

3- Immunology

4- Toxicology

5- Ethics of scientific research

6- Writing scientific research

ACKNOWLEDGMENT

Thanks my God, who gives me Knowledge and ability to finish this work.

I wish to express my deepest gratitude to:

Prof. Dr. Afaf Abd El- Galeel Abo Nour, Professor of Physiology, Zoology Department, Faculty of Women for Arts, Science & Education, Ain Shams University, for her supervision, kind encouragement, support in hard time and she suggested the plan to perfect work.

Prof. Dr. Adel Abdel-Aleem Shaheen, Professor of fish Diseases and Management, Department of Aquatic Animals Diseases and Management, Faculty Veterinary Medicine Benha University, for supervision, guidance, kind cooperation and his brilliant ideas to complete this work.

Assoc. Prof. Heba Salah Mohamed Hamed, Associate professor of Fish physiology, Zoology Department, Faculty of Women for Arts, Science & Education, Ain Shams University, much thanks from my heart for her great efforts, kind support, guidance, correction and her endurance for a long time.

My cordial thanks to my family for their support, kind encouragement donation and love.

ABSTRACT

Effect of *Nigella sativa* on silver nanoparticles induced hepato-nephrotoxicity in *Clarias gariepinus*

Silver nanoparticles (Ag-NPs) is effectively used in varied aquaculture applications. This study aimed to determine LC_{50} of (Ag-NPs) in African catfish *(Clarias gariepinus*) as well as the protective role of *Nigella sativa* (NS) against Ag-NPs toxicity. Fish were divided into 4 groups of triplicates as follows: group I: was showed as control, group II: was fed on 3% NS inclusion diet, group III: was exposed to 50 mg/L Ag-NPs and group IV exposed to 50 mg/L Ag-NPs and fed on 3% NS for 30 days.

At the end of experiment blood collected for biochemical analysis. As well as liver and kidney tissues of catfish were taken for oxidative enzymes, DNA analysis and histopatholgical studies.

The results exhibited that exposed of catfish with 50 mg/L Ag-NPs lead to a marked increase in liver enzymes Alkaline phosphatase (ALP), alanine amino transferase (ALT) and Aspartate amino transferase (AST). Also, these increases were indicated in kidney activities (urea, creatinine, and uric acid). Levels of total proteins, albumin and globulin were markdly increased. However, A/G ratio decreased. Also, glucose, cortisol and lactate dehydrogenase (LDH) levels were significantly increased. and total lipids levels (cholesterol and triglyceride) were increased. On the other and, African catfish exposed to Ag-NPs showed significant ($p<0.05$) elevation in the level of Lipid peroxidation (LPO) of liver and kidney tissues. These results indicated that the levels of reduction glutation (GSH), catalase (CAT), superoxide dismutase (SOD), and total antioxidant capacity (TAC) increased compared to the control fish after exposure to 50 mg/L Ag-NPs for 30 days.

Furthermore, percentage of DNA damage in both liver and kidney tissue of African catfish exposed to 50 mg/L Ag-NPs significantly increased.

Histological study in hepatic and renal tissues of African catfish represented more damage and alternation in tissues that exposed to 50 mg/L Ag-NPs for 30 days compare to control group and treatment group with 3% NS. It is concluded that the feeding of African catfish on a (NS) enriched diet minimized the destructive impacts of Ag-NPs toxicity and restored the above-mentioned variable near to the control group. The current investigation has revealed that dietary NS has antagonistic functions against Ag-NPs toxicity in African catfish.

Key words: Silver nanoparticles

Clarias gariepinus

Nigella Sativa

Biochemistry

DNA fragmentation

Histopathology.

CONTENTS

LIST OF TABLES

LIST OF FIGURES

	(G) with no inflammatory cellular infiltrate. The renal tubules lined by regular epithelium, slight dilaton of Bowman's space (B), melanomacrophages (Mm) and haemopoietic tissue (H&E, 400 x).	
25	Section in kidney tissue of African catfish *(C. gariepinus*) group (IV) which exposed to 50 mg/L Ag-NPs and fed on 3% NS showing patent glomeruli (G) with no inflammatory cellular infiltrate. The renal tubules lined by regular epithelium, melanomacrophages (Mm) and haemopoietic tissue (HT). (H&E, 400 x).	69

LIST OF ABBREVIATION

Ag-NPs	Silver nanoparticles
ALT	Alanine amino transferase
ALP	Alkaline phosphatase
Al_2O_3	Aluminum oxide
AST	Aspartate amino transferase
ATPase	Adenosine triphosphates
BM	Basemrnt membrane
B	Bowmen's space
BS	Blood sinusoids
BUN	Blood urea nitrogen
CAT	Catalase enzyme
$CaCO_3$	Calcium carbonate
CCL4	Carbon tetrachloride
CCG	Carbon coated copper grids
CPV	Cytoplasmic vaculation
CV	Central vein
D	Degenration
DFB	Diflubenzuron
DNA	Deoxyribonucleic acid
DTNB	Ditiobis- (nitro benzoate)
DHBS	5,5: dithiobis (2- nitro benzoic acid)
EDTA	Ethylene Diamine Tetra Acetic acid
G	Glomeruli
GSH	Reduced glutathione
g/l	Gram per liter
g/dl	Gram per deciliter
gcc^{-1}	Gram per cubic centimetre
GFR	Glomerular filtration rate
H	Hepatocytes
HT	Haemopoietic tissue
HRP	peroxidase
H_2O_2	Hydrogen per oxide
HCL	Hydrochloric acid

H&E	Hematoxylin and Eosin
IF	Inflammatory cells
Khz	Kilohertz
Kv	Kilovolt
LC_{50}	Half lethal concentration
LDH	Lactate dehydrogenase
LD_{50}	Half lethal dose
LPO	Lipid peroxidation
Mm	Melanomacrophage
Mg/dl	Milligram per deciliter
Mg/L	Milligram per liter
mRNA	Messenger ribonucleic acid
MDA	Malondialdehyde
MCP	Monocrotophos
µ/L	Micro per liter
µg/L	Microgram per liter
µg/dl	Microgram per deciliter
µg/Kg	Microgram per kilogram
µg/mg	Microgram per milligram
µm	Micrometer
µmol/mg	Micromole per milligram
N	Necrosis
NP	Nanoparticle
NS	*Nigella sativa*
nm	Nanometer
nmol/g	Nano moles per gram
nmol/mg	Nano moles per milligram
PM	Postmortem
PPm	Part per million
RC	Renal corpuscle
RCV	Rupture of the central vein
rpm	Revolution per minute
RBCs	Red blood cells
SOD	Superoxide dismutase
TAC	Total Anti-oxidant Capacity
TBA	Thiobarbituric acid

TEM	transmission electron microscopy
TG	Triglycerides
TiO_2	Titanium dioxide
W	Watt
ZnO	Zinc oxide
ROS	Reactive oxygen species

INTRODUCTION

Nanotechnology is considered as one of the key technologies of the 21st century and promises a revolution in our world **(Taju *et al.*, 2014).** Among the nanomaterials, silver nanoparticles are the most rapidly rising class with 320 t of production per year due to unique properties and consumer's weights **(Nowack *et al.*, 2011).** Silver nanoparticles are manufactured for their antimicrobial properties **(Dos Santos *et al.*, 2014 and Bruneau *et al.*, 2016).** They have been shown to be effective mainly as an antibacterial against both Gram-positive and Gram-negative bacteria aquatic pathogenic bacteria. They are used extensively in clothing, baby products (e.g. nipples and bottles), personal care products (e.g. shampoos, toothpastes, deodorants, cosmetics, soaps, and food containers (**Ribeiro *et al.*, 2014, Taju *et al.*, 2014 and Bruneau *et al.*, 2016).** As weell as, bedding and appliances (e.g. washing machines, humidifiers and refrigerators) **(Woodrow Wilson International Center for Scholars, 2008).**

The antimicrobial properties of silver nanoparticles (Ag-NPs), with a high surface area ratio, could modify Ag toxicity and class them in a specific contaminant category **(NRC, 2012).** The discharge of silver nanoparticles into aquatic and terrestrial environment as a result of the extensive use of Ag nanoproducts increased through weathering of rocks, burning of fossil fuel, processing of ores, cement manufacturing, leaching and anthropogenic activities **(Awasthi *et al.*, 2013 and Taju *et al.*, 2014).** Rain is responsible to release the silver in the ground and water reservoirs **(Wijnhoven *et al.*, 2009).** Several investigations studied the toxicological action of Ag NPs on fishes **(Choi *et al.*, 2010 and Lee *et al.*, 2012),** where the effect levels of Ag-NPs ranged from μgl^{-1} to mgl^{-1} including the variety of aquatic organism type, particle type, and exposure conditions **(Farmen *et al.*, 2012)**. Exposure to Ag-NPs caused cytotoxicity, oxidative stress, increased lipid peroxidation, reduced mitochondrial function, genotoxicity **(Scown *et al.*, 2010 and Bruneau *et al.*, 2016)**, apoptosis **(Sayed, 2016)**, protein degradation and DNA damage **(Gagné *et al.*, 2013)**. Also, they caused depletion of oxidative stress markers **(Arora *et al.*, 2009)**. Also, Ag-NPs can bind onto cell membranes and affect permeability and respiratory function of the cells by decreasing the Na+/K+ and ATP activity **(Morones *et al.*, 2005)**.

Nigella sativa (NS), or also known as black cumin or its Arabic name habat-ul sauda, has been used for centuries a medicinal plant and as a food additive, throughout the Middle East, India, North Africa and Asia **(Shahid *et al.*, 2017)**. It is an annual flowering plant with dark-crescent shaped seeds and belongs to Ranunculaceae family. The seeds are considered to be the most valuable part contributing beneficial health effects **(Terzi *et al.*, 2010)**. NS as a natural remedy has been documented to possess numerous therapeutic values, including diabetes, tumour, hypercholesterolemia, hypertension **(Taşar *et al.*, 2012 and Leong *et al.*, 2013)**. It also acts as anti-inflammatory, antihypertensive, antimicrobial, hypolipidemic, antioxidant, hypoglycemic, anticytotoxic and neuroprotective effects **(Kanter *et al.*, 2006 and Amin and Hosseinzadeh, 2016)**. In addition, the protective effect of NS against nephrotoxicity induced by Cyclosporine **(Shahid *et al.*, 2017)**.

African catfish *(C. gariepinus)* is one of the most important tropical cultured fish due to high growth rate, high stocking density capacities and high resistance to poor water quality and oxygen depletion **(Karami *et al.*, 2010; Marzouk *et al.*, 2017)**.

Due to it's widely distribution throughout Africa (**Marzouk *et al.*, 2017**). It is used in fundamental research as an excellent model for toxicological studies (**Mekkawy *et al.*, 2011**). In conclution, the present study was undertaken to elucidate the protective role of dietary supplementation of NS seeads against Ag-NP induced. Biochemical alteration, of liver and kidney oxidative stress *via* determination of lipid peroxidation and endogenous antioxidants status. Also, induced DNA damage, histopathological alterations

AIM OF THE WORK

The present study was planned to investigate the protective effect of *Nigella sativa* (NS) against the effect of silver nanoparticles (Ag-NPs) in African catfish *(C. gariepinus)*. This study aims also to evaluate the effects of Ag-NPs on the levels of the biochemical parameters (Liver and kidney functions), and serum immune function (serum total proteins, serum albumin, serum globulin and A/G ratio). Also, indicated its effect on glucose, cortisol, lactate dehydrogenase and total lipids as well as the levels of antioxidant and oxidative stress (LPO, GSH, CAT, SOD and TAC). In addition, the present study aims also to investigating the effect of (Ag-NPs) and the protection by *Nigella sativa* on the DNA fragmentation.

Moreover, this study was carried out to evaluate the histopathological change in exposed *C. gariepinus* to Ag-NPs. Finally, this study aims also to represent the effect of NS to improvement all the above changes.

REVIEW OF THE LITERATURE

Sliver nanoparticles (Ag-NPs):

Silver nanoparticles are between 1 nm and 100 nm in size. Used Ag-NPs are spherical, but octagonal and diamond. Ag-NPs have unique optical and chemical properties

Ag-NPs and half lethal concentration (LC_{50}):

Silver nanoparticles has toxic effect so its important to determine LC_{50}. The 96 h LC_{50} tests is method for finding the toxicity of more elements. In **(2009) Bar-Ilan *et al*.** recorded that Ag-NPs in zebra fish (*Danio rerio*) calculated LC_{50} values were 93.31 μm for 3nm particle size and 137.26 μm for 100 nm.

According to **Johari *et al.* (2013)** calculated the LC_{50} of Ag-NPs in rainbow trout *(Oncorhynchus mykiss)* values were 0.025 mgl^{-1}, 0.71 mgl^{-1} and 2.16 mgl^{-1} for Eleutheroembryo, Larvae and Juveniles. In the same year, **Shaluei *et al.* (2013)** showed LC_{50} Ag-NPs value in silver carp (*Hypophthalmicthys molitrix*) calculated (0.810, 0.64, 0.383 and 0.202 mgl^{-1} for 24, 48, 72 and 96h.

By **Yalsuyi and Vajargah. (2017)** determined that the toxicity of silver nanoparticles on goldfish (*Carassius auratus* gibelio) for 24, 48, 72 and 96 h. the results revealed that the half lethal value of silver was 27/017, 20/245, 14/558 and 9/781 respectively. While in **(2018) Kumar *et al.*** indicated that the (LC_{50}) values of silver nanoparticles in fish Murrel *(Channa striatus*) was 1/25 (4 ppm) for 96 h. of exposed. Also, **pandit and Sinha (2019)** recorded that the (LC_{50}) of silver nanoparticles for Indian air-breathing *Clarias batrachus* (L.) for 24, 48, 72 and 96 h. the results revealed that the half lethal values of silver were 346.20, 237.33, 201.28 and 171.16 mg/L. Respectively

Ag-NPs and clinical signs and behavioral effects:

Previous clinical studies investigating toxic effect of Ag-NPs on catfish. **Üral and Koprucu (2006)** found erratic swimming and swimming at the water surface then noted rapid gill movement in European catfish (*Silurus glanis)* exposed to dichlorvos higher than 8 mg/L. According to **Kavitha and Rao (2007)** observed loss of equilibrium, erratic swimming and loss of feeding in mosquito fish (*Gambusia affinis*) exposed to (20.49 mg/L.) monocrotophos for 96h.

Sayed (2016) recorded an abnormal behavior in African catfish *(C. gariepinus*) that exposed to 50 mgl^{-1} and 75 mgl^{-1} silver nanoparticles for 14 days compare to control group such as quick swimming movement on the back, rolling movements plus appearance of pigment skin. Also, study by **Sayed and Soliman (2017)** found abnormalities included notochord curvate (C-shape curvature) and yolk sac edema in catfish *(C. gariepinus)* after exposed to (30 h-PFS, 75 ng/ L) Ag-NPs. In the same year, **Hamed and Osman (2017)** noted loss equilibrium were followed handing vertically in the water and irregular erratic swimming and observed convulsion, seizures and increase opercula movement after exposure to carbofuran for one month in African catfish *(C. gariepinus)*.

Recently, **Naguib *et al.* (2020)** observed that African catfish *(C. gariepinus*) has abnormal behavior such as fast swimming, loss of equilibrium and change in color skin this occur when exposed to 100 mg/L Ag-NPs for 15 days.

Ag-NPs and liver function:

Lee *et al.* (2012) noted that exposure of (200 mg/L) higher concentration of Ag-NPs caused increase ALP serum level in common carp (*Cyprinus carpio)*. In the same line, **Rajkumar *et al.* (2015)** recorded that ALP level increase in Rohu (*Labeo rohita)* after exposed to Ag-NPs. And **Abdel-wahhab *et al.* (2016)** showed that serum level of ALT, and AST increased in *C. gariepinus* treated with ochratoxin A (OTA) for 21 days.

Monfared and Soltani (2013) observed a significant increase of ALP, ALT and AST levels after exposed to Ag-NPs in rainbow trout (*O. mykiss*). Moreover, **Abdel-Khalek *et al.* (2015)** showed a significant increase in ALP, ALT and AST levels of Nile Tilapia (*Oreochromis niloticus)* after exposed to copper oxide nanoparticles. In the same year, **Imani *et al.* (2015)** noted a significant increase in the activities of ALP, ALT and AST of rainbow trout (*O. mykiss*) after exposed to Ag-NPs. In this concern mention has been recorded by **Abdel-Tawwab and Hamed (2018)** that ALP, ALT and AST values in Nile tilapia *(O. niloticus*) increase significantly after exposure to 1.64, or 3.28 μg BPA/L for 6 weeks. Also, **Kumar *et al.* (2018)** observed an increase in the levels of ALT and AST in *Channa striatus* after exposed to Ag-NPs. **Kovacik *et al.* (2019)** noted a significant increase AST serum level in freshwater fish *(Cyprinus carpio*) after exposed to trace metals. And **Naguib *et al.* (2020)** observed a significant increase in ALP, ALT and AST in African catfish *(C. gariepinus)* after exposed to Ag-NPs for 15 days.

In contrast, **Sayed and Hamed (2017)** noted decrease in ALP level in exposed group 4-NP for 14 days African catfish *(C. gariepinus*), and noted increase in ALT and AST level. **Hedayati *et al.* (2019)** marked a significant decrease in ALP, ALT and AST in zebrafish (*D. rerio*) after exposed to silver nanoparticles (Ag-NPs) and silver nitrate ($AgNO_3$).

Ag-NPs and Kidney function:

In (2006), Radwan and El-Said recorded a marked effect of two commercial formulations of profenofos (0.566 and 0.850 mg/L of selecron and 0.234 and 0.390 mg/L of cord) for one and two weeks in Nile tilapia fish (*O. niloticus)*, caused increase in urea and creatinine levels. Other investigators, **Abdel-Khalek *et al.* (2015) and Alkaladi *et al.* (2015)** recorded a highly significant increase in serum creatinine and uric acid level of *O. niloticus* after exposure to zinc and copper oxide nanoparticles. And **Abdel-wahhab *et al.* (2016)** showed increase in serum level of creatinine and, uric acid after OTA exposure in *C. gariepinus*.

Sayed and Hamed (2017) recorded that urea, creatnine and uric acid were significantly increase in African catfish *(C. gariepinus)* which exposed to 4-nonylphenol. In the same year, **Hamed and Osman (2017)** noted increase serum urea and creatinine in African catfish *(C. gariepinus*) after exposed to carbofuran for 4 weeks. In this concern, **Abdel-Tawwab and Hamed (2018)** observed an increase in creatinine and uric acid in Nile tilapia fish, (*O. niloticus*) exposed to 3.28µg/L BPA/L for 6 weeks. In the same trained, **Mahmoud *et al.* (2019)** showed increase in serum creatinine and uric acid in African catfish *(C. gariepinus*) after exposure to Ag-NPs for 15 days.

Ag-NPs and Total proteins and immune function:

According to **Öner *et al.* (2008)** demonstrated a significant increase in total protein level in freshwater Nile tilapia fish (*O. niloticus*) after exposure to Ag (4.35±0.25 mg/L.). whenever

Monfared *et al.* (2013) showed that exposure to Ag-NPs in rainbow trout *(O. mykiss*) caused decrease in total protein level.

In (2015), Atli *et al.* observed effect of acute 10 μM and cronic 20 Mm exposure of Cd and Pb in Nile tilapia fish (*O. niloticus*) lead to no change in serum total protein level. **Sayed *et al.* (2016)** noted decrease in total protein in African catfish *(C. gariepinus*) after exposed to 4-NP (100μg/L). in addition to **Hamed and Osman (2017)** showed the effect of carbofuran in African catfish *(C. gariepinus)* noted decrease in total proteins serum level after 4 weeks. **In (2016), Abdel-wahhab *et al.*** Showed an increase in albumin, globulin and A/G ratio in group exposed to chitosan nanoparticles in African catfish (*C. gariepinus*). In the same idea, **Hamed and Abdel-Tawwab (2017)** recorded a decrease in total proteins, albumin and globulin levels in Nile tilapia (*O. niloticus)*, induced by exposure to BPA for 6 weeks.

Abdel-Tawwab and Hamed, 2018 noted that level of total protein, Albumin and globulin were decreased in Nile tilapia fish (*O. niloticus*) due to its exposed to 3.28 μg/L BPA/L. On other hand, **Vajargah *et al.* (2019)** recorded the effect of Ag-NPs in male goldfish (*C. auratus gibelio*) showed a significant increase in serum total protein. Recently, **Abd El-Atti *et al.* (2020)** noted a decrease in the total protein of the crayfish (*procambarus clarkii*) after exposed to 125 and 250 ppm of Ag-NPs for 28 days.

Ag-NPs and Stress indicators:

According to **Monfared *et al.* (2013)** noted that exposure to Ag-NPs in rainbow trout (*O. mykiss*) caused elevation in glucose level. Other results by **El-Bahr and Azraqi (2014)** illustrated that effects of black cumin on blood glucose was over flow and reduce total value in serum level. Moreover, **Abdel-Khalek *et al.* (2015)**

observed a significant increase in serum glucose of Nile tilapia (*O. niloticus)* after exposure to copper oxide nanoparticles. Similar results were reported by **Hamed and Osman (2017)** who observed an increase in serum glucose level after exposed to carbofuran in African catfish (*C. gariepinus)* for 4 weeks. These results agree with **Sayed and Hamed (2017)** noted that exposed to 4-NP increase glucose level in African catfish (*C. gariepinus)*. And **Kumar *et al.* (2018)** showed that level of glucose serum increase in *C. striatus* after exposed to Ag-NPs.

In the same line, **Kovacik *et al.* (2019)** recorded increase glucose serum level in Common carp (*C. carpio*) after exposed to trace metals. By **Vajargah *et al.* (2019)** recorded significant change in serum glucose after exposed to Ag-NPs in male goldfish (*Carassius auratus gibelio*). In the same year, **Hedayati *et al.* (2019)** marked higher in 1.5+0, 200+0, 200+1.5 and 400+0 groups serum glucose in zebrafish (*D. rerio)* after exposed to silver nanoparticles (Ag-NPs) and silver nitrate ($AgNO_3$). **Naguib *et al.* (2020)** showed significant increase in serum glucose level in African catfish (*C. gariepinus*) exposed to Ag-NPs for 15 days.

In other results **Monfared *et al.* (2013)** observed that exposure to Ag-NPs in rainbow trout (*O. mykiss*) caused increase in cortisol level. And **Sayed *et al.* (2016)** noted increase in cortisol level after exposure to100 µg/L 4-nonylphenol in African catfish *(C. gariepinus)*. These results suitable with **Hamed and Osman (2017)** who recorded a significant increase in serum cortisol level after exposed to carbofuran in African catfish *(C. gariepinus)* for 30 days. Also, **Kumar *et al.* (2018)** observed that level of cortisol increase in *C. striatus* after exposed to Ag-NPs.

Ag-NPs and Lactate dehydrogenase enzyme (LDH):

According to **Wu and Zhou (2013)** observed that the decrease in LDH level activity in *Oryzias latipes* because of exposed to silver nanoparticles after 14 days. This contrast to **Atli *et al.* (2015)** showed effect of (acute 10 μM and cronic 20 μM) exposed Cd and Pb in *O. niloticus* lead to increase in LDH level of fish exposed chronic Pb and variation in exposed to Cd.

In **(2016), Sayed *et al.*** noted that LDH level decreased in African catfish *(C. gariepinus*) after exposed to 4-Np. These agree with **Abhijith *et al.* (2016)** marked that LDH level decreased in liver of catla carp (*Catla catla*) after exposure to methyl parathion. In the same line, **Sayed and Soliman (2017)** recorded that LDH level decrease in embryos of African catfish (*C. gariepinus)* after exposed to silver nanoparticles (Ag-NPs) from 24 to 144 h-PES. In addition, in **(2018)** the same author showed that LDH level decreased after exposure to 4-NP in African catfish (*C. gariepinus)*.

In the same year, **Kumar *et al.* (2018)** observed that level of LDH increase in *C. striatus* after exposed to Ag-NPs. But **Hedayati *et al.* (2019)** found a marked increase in LDH level in Zebrafish (*D. rerio*) after exposed to silver nanoparticle (Ag-NPs) and Silver nitrate ($AgNO_3$). Recently, **Abd El-Atti *et al.* (2020)** found a significant increase in the level of LDH in hepato-pancreas of crayfish (*P. clarkii*) exposed to 250 ppm of Ag-NPs for 28 days.

Ag-NPs and lipids parameters:

A study was conducted by **Yousafzai and Shakoori (2011)** showed increase serum total lipids and cholesterol in freshwater fish against aquatic pollution. This agree with **Valerio-Garcíaa *et***

***al.* (2017)** who recorded a significant increase in the total lipids of *Chapalichthys pardalis* after exposed to Ag-NPs.

Recently **Abd El-Atti *et al.* (2020)** recorded increase in serum total lipid in crayfish (*P. clarkii*) after exposed to 125 and 250 ppm of Ag-NPs for 28 days. In the same year, **Naguib *et al.* (2020)** reported a significant increase in serum total lipid in African catfish *(C. gariepinus*) after exposed to Ag-NPs.

In (2017), Sayed and Hamed noted an increase in cholesterol after exposed to the 4-NP in African catfish *(C. gariepinus*). In addition, **Vajargah *et al.* (2019)** recorded the effect of Ag-NPs in male goldfish (*C. auratus gibelio*) observed a significant increase in cholesterol serum level but did not change in triglyceride. In the same year, **Kovacik *et al.* (2019)** showed a significant increase cholesterol level in *Cyprinus carpio* after exposed to trace metals. A study had been mentioned by **Öner *et al.* (2008)** mention that increase serum triglyceride in Nile tilapia (*O. niloticus*) caused by exposure to Ag with the high level (842.2 ± 62.8 mg/L.).

<u>Ag-NPs and Oxidative stress and Anti-oxidants enzymes:</u>

lipid peroxidation (LPO):

According to **Strużyński *et al.* (2013)** observed increase LPO level in spinycheek crayfish (*O. limosus*) after exposure to silver nanoparticles. Other results by **Wu and Zhou (2013)** Observed that increase LPO level activity in *Oryzias latipes* because of exposed to silver nanoparticles (Ag-NPs) after 14 days. Similarly, **Benavides *et al.* (2016)** showed a significant increase hepatic (LPO) level in *Carassius auratusafter* exposed to aluminum

(Al_2O_3) and (ZnO) oxide nanoparticles for one and two weeks. This agree with **Sayed and Soliman (2017)** recorded an increase in LPO level after exposed to silver nanoparticles (Ag-NPs) from 24 to 144 h-PES in African catfish (*C. garipinus*). It has been confirmed by **Zebral *et al.* (2019)** mentioned that hepatic LPO level in killifish (*Poecilia vivipara*) was elevated at 28 °C exposed to 9 µg/ L or 20 µg/L of copper.

Recently, **Abd El-Atti *et al.* (2020)** showed that crayfish (*P. clarkii*) exposed to Ag-NP increased in LPO level after 28 days. In the same year, **Faheem *et al.* (2020)** noted an increase in the LPO level after exposure to BPA concentrations (100, 1000 µg/l) in Larvae of *Labeo rohita* fish.

Glutathione content (GSH):

According to **Jin *et al.* (2010)** demonstrated that GSH was decreased in zebrafish (*D. rerio*) exposed to (1, 10, 100 and 1000 µg/l) atrazine for 2 weeks. And **Xiong *et al.* (2011)** showed that GSH concentration were decreased in liver tissue after exposure to 50 mg/L of TiO_2 in dark, 5 mg/L of ZnO NPs and bulk ZnO suspensions for 96 h. on zebrafish (*D. rerio*). **In (2015), Kaya and Akbulut.** observed the effect of 1 mg/l zinc oxide (ZnO) nanoparticles on Nile tilapia (*O. niloticus*) for two weeks recorded that a significant decrease in GSH level. In the same year, **Rajkumar *et al.* (2015)** showed GSH level which reduction after exposed to Ag-NPs in *L. rohita*.

On the other side **Yuan *et al.* (2016)** showed a significant increase in GSH level after exposed to bisphenol in rare minnow *(Gobiocypris rarus*). Recently, **Abd El-Atti *et al.* (2020)** marked that crayfish (*P. clarkii*) exposed to 125 and 250 ppm Ag-NPs increase in GSH level after 28 days but not change in 25 ppm Ag-NPs.

Catalase enzyme (CAT):

According to **Lee *et al.* (2012)** observed that exposure of (200 mg/L) Ag-NPs caused increase CAT activity in liver of common carp (*C. carpio*). But results by **Wu and Zhou (2013)** showed that CAT levels were decrease activity in *Oryzias latipes* because of exposed to silver nanoparticles (Ag-NPs) after 14 days.

In **(2014), Dalibor *et al.*** reported increase in the CAT level of signal catfish (*Pacifastacus leniusculus*) exposed to triazine for 30 days. In the same line, **Jayaseelan *et al.* (2014)** recorded a significant increase activity of CAT in Mozambique tilapia *(Oreochromis mossambicus)* after exposed to Nickel nanoparticles for 14 days.

Kaya and Akbulut (2015) observed a significant increase in CAT level after exposed to 10 mg/l zinc oxide nanoparticles on Nile tilapia (*O. niloticus*) for 7 days. A study by **Sayed *et al.* (2016)** show a significantly increase in CAT level in African catfish *(C. gariepinus)* after exposed to 4-nonylphenol (100 μg/L). Recently, **Abd El-Atti *et al.* (2020)** noted that crayfish (*P. clarkii*) exposed to 25 ppm Ag-NPs not change in the level of CAT but increase in 125 and 250 ppm for 28 days.

Superoxide dismutase (SOD):

In **(2013), Wu and Zhou** Showed that decrease SOD level in *Oryzias latipes* because of exposed to silver nanoparticles (Ag-NPs) after 14 days. Other results in **Jayaseelan *et al.* (2014)** recorded a significant increase in SOD activity after exposed to Nickel nanoparticles for 14 days in Mozambique tilapia (*O.*

mossambicus). By **Kaya and Akbulut (2015)** showed the effect of 1 mg/l zinc oxide (ZnO) nanoparticles on Nile tilapia (*O. niloticus*) for 7 days recorded that a significant increase in SOD level. In the same year, **Rajkumar *et al.* (2015)** recorded SOD level which decrease after exposed to Ag-NPs in Rohu (*labeo rohita*).

Benavides *et al.* (2016) noted a significant increase SOD level in freshwater fish (*C. auratusafter*) exposed to aluminum (Al2O3) and (ZnO) oxide nanoparticles for 7 and 14 days. In the same line, **Sayed *et al.* (2016)** noted an increase in SOD level of catfish *(C. gariepinus*) after exposed to 4-nonylphenol (100μg/L). Moreover, **Hamed and Osman (2017)** studied increase in SOD level in liver and kidney African catfish *(C. gariepinus*) after exposed to carbofuran for 30 days. Recently, **Faheem *et al.* (2020)** observed a decrease in SOD level in *Labeo rohita* larvae after exposure to BPA concentrations for 21 days.

Total antioxidant capacity (TAC):

Study was conducted by **Abdel-wahhab *et al.* (2016)** noted an increase in total antioxidant capacity (TAC) level in liver and kidney of African catfish *(C. gariepinus*) after exposed to Chitosan nanoparticles. This results contrast with **Sayed *et al.* (2016)** recorded that TAC Level decreased after exposed to 4-nonylphenol (100 μg/L) in African catfish *(C. gariepinus).*

Similarly, **Abou Khalil *et al.* (2017)** stated that level of TAC increased in African catfish (*C. gariepinus*) after exposer to (25 and 50 g/kg) 4-nonylphenol. This agree with **Qiu *et al.* (2019)** showed an increase in TAC level after exposed to 1000 mg/L BPS for 60 days in common carp fish (*C. carpio*). In the same year, **Zebral *et al.* (2019)** observed that hepatic TAC level was elevated at 28 °C after exposed to 9 µg/L of copper in killifish (*P. vivipara*).

Ag-NPs and DNA fragmentation analysis:

According to **Rothkamm and Lobrich (2003)** noted that exposed to Ag-NPs in zebrafish (*D. rerio*) causes DNA damage includes double strand breaks cause lesions in cells. In **(2010), Park and Choi** recorded increase in DNA strand breaks which caused legend in cells in *Daphina Magana* after exposed to 1 mg/L Ag-NPs. Moreover, **Gagńe *et al.* (2013)** observed a formation of DNA strand breaks in freshwater mussels (*Elliptio complanata*) which exposed to silver nanoparticles (Ag-NPs) for 48 h at 15˚C.

Massarsky *et al.* (2014) noted that exposure of Ag-NPs in rainbow trout (*O. mykiss*) caused increase the DNA damage. In the same year, **Taju *et al.* (2014)** recorded that exposed to different concentration (2, 4, 8, 16, 32 and 64 µg/ml) of (Ag-NPs) make DNA damage of *Catla catla* and *Labeo rohia.* A study was conducted by **Sayed, (2016)** showed that highest percentage of DNA damage in liver and kidney appearance of apoptic cells. in African catfish *(C. gariepinus*) which exposed to different concentration (25, 50 and 75 mgl^{-1}) Ag-NPs, respectively. Other results in **Sayed and Hamed (2017)** recorded that the percentage of DNA fragmentation in African catfish (*C. garieinus*) was increase after exposure to 4-nonylphenol (0.1mg/L) for 14 days.

Sayed and Soliman (2018) marked that African catfish *(C. gariepinus*) increased in percentages of DNA fragmentation after exposed to 4-nonylphenol. Recently, **Abd El-Atti *et al.* (2020)** observed increase in the damage of DNA in crayfish (*P. clarkii*) exposed to Ag-NPs 250 ppm for 28 days. In the same year, **Liu *et***

***al.* (2020)** noted increase DNA damage in rats after exposed to Ag-NPs.

Ag-NPs and histopathological change:

In **(2006), Gonzalaz *et al.*** noted that the accumulated Ag-NPs in zebra fish (*D. rerio*) cause damage in the liver, the alterations are hepatic cell cord, condensation of chromatin, apoptic change and pyknosis in adult. By **Choi *et al.* (2010)** regarded a number of cellular alterations (disruption of hepatic cell cords, apoptotic changes such as pyknosis and chromatin condensation) in the liver of adult zebrafish which exposed to 120 mg Ag/L, Ag-NPs for 24h.

In addition, **Wu and Zhou (2013)** recorded histological change in liver and kidney tissue such as hepatocyte enlargement, global basophilic and hepatic impairments in *Oryzias latipes* after 14 days of exposure of silver nanoparticles. In the same trend, **Lee *et al.* (2012)** demonstrated that after exposed common carp (*C. carpio*) to 100 mg/L Ag-NPs induced histological change in liver due to accumulation of eosinophilic granules and atrophy of hepatocyte nucleus, after exposed to 200 mg/L Ag-NPs caused accumulation of eosinophilic granules.

Other investigators, **Benavides *et al.* (2016)** Showed that goldfish (*C. auratus gibelio*) exposed to aluminum (Al_2O_3) and (ZnO) oxide nanoparticles for 14 and 21 days presented clear degeneration of hepatic tissue. Other results such as **Hamed and Osman (2017)** founded inflammatory cells, largamente of the left lobe of the liver, necrosis, hemorrhage and vacuolization after exposed to carbofuran for 30 days in African catfish *(C. gariepinus)*.

Moreover, **Ostaszewska *et al.* (2018)** observed that Ag-NPs caused many histological abnormalities in the liver nuclear degradation, necrotic cell with dilated endoplasmic reticulum, cell degeneration and macrophage with nanoparticles numerous mitochondria and peroxisomes of rainbow trout (*O. mykiss*) after 28 days of exposed to 1.5 mgL^{-1} of silver nanoparticles.

Recently, **Naguib *et al.* (2020)** recorded that exposure of *C. gariepinus* to 100 mg/l of silver nitrate ($AgNO_3$) for 15 days caused hepatocytes delimited by ruptured cell membrane, some hepatocytes loss nuclei while other were pyknotic nuclei. Whereas, in 10 mg/l of Ag-NPs for 15 days in African catfish *(C. gariepinus)* who noted more damage; hepatocytes lost their normal polygonal shape. Hepatic cells decreased in size and found with no nucleus while other had pyknotic nuclei.

On the other hand, **Kotb *et al.* (2018)** reported that histological alterations of kidney of African catfish *(C. gariepinus*) has change in glomerular cell, reduction in the size of the glomular vessels, lowering nephron expression and Bowman's space due to accumulated edema, increase fluid movements from glomular pool, thin basement membrane and increase fenestrae in between glomerular vascular endothelium after exposed to 4-nonylphenol.

Nigella sativa:

Nigella sativa or black cumin used for treatment and have beneficial health effect this agree with **Al-Jishi (2000)** observed that black cumin NS used in medicine for treatment such as rheumatic diseases, common cold, fever and asthma, it is used for good health as a flavoring additive to prickles and bread. Addition to **Mohammed and Arias (2016)** recorded that NS seeds has strong protective effect in catfish, *Ictalurus punctatus* and *D. rerio*.

On the other hand, **Diab *et al.* (2004)** studied the effect of NS on serum total protein and albumin of *O. niloticus*. In the same line **Ezzat (2005)** demonstrated the effect of the NS on the total protein, albumin and globulin in serum of African catfish *(C. gariepinus*). This results suitable with **Al-Kubaisy and Al-Noaemi (2007)** evaluated the effect of *Nigella sativa* which decrease levels of protein and albumin after treatment with $CCl4$. Another study by **Hasan *et al.* (2018)** showed a reduction in ALT and AST serum level with *Nigella sativa* and it was noted that enzymes were affected by it. In **(2019), Aly *et al.*** demonstrated the effect of *Nigella sativa* on biochemical results such as ALT, AST, kidney function, total protein and globulin. **Latif *et al.* (2021)** demonstrated decreased liver and kidney biochemical parameters (ALP, ALT and AST) (urea and creatinine) levels in *L. rohita* fed black seed diet.

Mohammed (2003) recorded that NS is perfect source for protein and oil was rich in β-sitosterol that decrease the absorption of cholesterol. Other results by **El-Bahr and Azraqi (2014)** noted the effect of black cumin on blood glucose was overflow and reduce total cholesterol value in serum level in Streptozotocin induced Diabetic Rats. Other results showed by **Abdelrazek *et al.* 2018** investigated the effects of black seeds in diabetic rats in one month that reduced serum glucose level and lipid profile (Cholesterol and Triglycerides). **El –Far *et al.* (2017)** showed that NS seeds reduce total lipids in serum level and prevent the oxidative hepatic destroy in rabbit exposed to lead acetate.

By **Muhammad and Al-Ghamdi (2002)** investigated the effects of NS action include anti-oxidant and anti-inflammatory. This results supported by **Luther *et al.* (2007)** which noted the effect of black cumin oil in decrease antioxidant activities. **Abdelrazek *et al.* (2018)** recorded increase in GSH activities in diabetic rats for one month. And **Latif *et al.* (2021)** observed the effect of *Nigella sativa* supplementation on oxidative stress, result showed decreased lipid peroxidation levels but increased CAT level. In the same year, (**Mohmoud *et al.*, 2021)** recorded the effect of NS seeds on serum catalase activity (CAT) and (TAC) activities were increased in rats of 30 and 50 g/kg compared to control group for 30 days.

There is growing insert in the role of NS recorded by **Ezzat (2005)** studied the effect of the NS oil which decreased hepatic degeneration, necrosis and hepatoportal congestion in African catfish (*C. gariepinus)*. Moreover, **Kotb *et al.* (2018)** reported that *Nigella sativa* has effect on kidney such as decrease the glomerular cellular, reduction in the size of the glomerular vessels and lowering nephron expression in African catfish (*C. gariepinus)* which exposed to 4-nonylphenol. And **Abd-Elkareem *et al.* (2020)** observed that NS seeds have protective role in DNA damge in the African catfish (*C. gariepinus*) and protected the renal tubular basement membrane from 4-NP induced damage.

MATERIALS AND METHODS

Materials:

Experimental fish:

In the present study, 180 African catfish (*C. gariepinus*) were obtained from Abassa fish farm, El sharkaya governorate, Egypt. the average body weight was 300 ± 20 g. The fish were apparently healthy and free from any visible lesions. Fish were transported to fish disease laboratory, at the department of aquatic animals disease and management Faculty of Veterinary Medicine, Benha University in large plastic containers supplemented with battery aerators as a source of oxygen. In the laboratory, fish were acclimatized in the glass aquaria for two weeks, prior to the experiments.

Management:

Fish:

Glass aquaria measuring (60x47x32 cm) of 60-litre capacity were used, ten fish were kept in each glass aquarium containing dechlorinated tape water. The aquaria were cleaned and water was changed every 48 hours for experimental and control group. The aquaria were supplied with electric air pumps where dissolved oxygen was maintained at 6±0.5 ppm. All aquaria pH ranged from 7.4 ±0.3 by regulator water change, water temperature was adjusted at 28±2°C, alkalinity 122 mg/l and total hardness was 152 mg/l $CaCO_3$.

Chemicals:

Silver nanoparticles (Ag-NPs):

The Ag-NPs were purchased from Biodiagnostic Trade Co., Egypt.

Other chemicals used:

Potassium permanganate: used for treating of fish skin ulcer and cleaning. Were purchased from Stanbio Company, Cairo, Egypt.

Sodium thiosulphate: used for dechlorinization of tape water. Were purchased from Stanbio Company, Cairo, Egypt.

Chemical kits for biochemical determination such as: ALP, ALT, AST, urea, creatinine, uric acid, total protein, albumin, globulinglucose, cortisol, LDH and Total lipids were purchased from Stanbio Company, Cairo, Egypt.

Chemicals for antioxidant determination (LPO, GSH, CAT, SOD and TAC) were purchased from Bio-diagnostic Company, Cairo, Egypt.

Materials: used in blood sampling, DNA fragmentation and histopathological examination studies.

Methods:

Preparation and characterization of silver nanoparticles Ag-NPs:

Physical characteristics of the particles according to the manufactures data were; purity (99.5 %), trace metal basis, size ($\leq$100 nm) and density (10.49 gcc^{-1}). According to **Ghosh *et al.* (2012),** Ag-NPs were suspended in PBS and dispersed by ultrasonic vibrations (100 w 30 khz) for 30 min. The concentration of the Ag-NPs stock after filtration was 3975 mg^{-1}, that was very close to the concentration provided by the manufacturer (4000 mg^{-1}) measured by spectrometer. The stock suspension of Ag-NPs was resolicited for 3 to 5 min before tests, and diluted with deionized water to make final test concentration.

For TEM analysis, a drop of the solution (10μl) was placed on the carbon coated copper grids (CCG) and dried by allowing water to evaporate at room temperature. Electron micrographs were obtained using JEOL GEM-1010 transmission electron microscope at 70 kV, The Regional Centre for Mycology and Biotechnology, Al- Azhar University.

Determination of 96 hours half lethal concentration (LC_{50}) of silver nanoparticles:

48 catfish were used in different Six concentrations of Ag-NPs to determine the LC_{50}. The concentration which caused 50 % mortality as (LC_{50}) for 96 h of exposure were calculated according to **Litchfield and Wilcoxon (1949).** Eight fish were used for each concentration.

Half lethal concentration of Ag-NPs = Highest con. – Σ a x b /n

LC_{50} = 656 – 3875/8 = **171.7 mg/L**.

Where: n: Number of fish in each group.

a: Constant factor of difference between groups.

b: Mean value of dead fish between each two successive groups.

Preparation of diet:

Fish were fed twice daily on a commercial balanced fish diet were prepared according to **(Craig *et al.*, 2017).**

Nigella sativa (black seed) was purchased from local market in a powder form and mixed 3% of the diet is weight. Batches of feed were left for 24hr for air-drying and stored in 4°C for daily use **(Salah *et al.*, 2019 and AVMA, 2017).**

Experimental design:

120 catfish were divided into 4 groups, and each group consisted of 3 replicates of 10 fish per aquarium and fed on a diet which was given to fish up to satiation daily at 9:00 and 14:00 pm for 4 weeks. Group I: was served as control group without any treatment. Group II: was served as control group and co-administrated with 3% NS. Group III: was exposed to 50 mg/L Ag-NPs in water was prepared according to **Ghosh *et al.* (2012).** Group IV: was exposed to 50 mg/L Ag-NPs and fed on 3% NS. During the experimental period, any fish mortalities or abnormalities were recorded.

Clinical investigation and post mortem examination:

The experimentally exposed fish were kept under proper observation during the period of experiment for any clinical abnormalities, postmortem (PM) lesions or deaths. according to the method described by **Amlacher (1970).**

Sampling serum:

After 30 days, of exposure to Ag-NPs at the end of experiment:

1- Blood samples: were collected from the caudal vessels using plastic syringes in dry sterilized vials. Allowed the blood to clot then samples were centrifuged, at 3000×g at 4° C for 15 min. serum collected and stored at -20° C until used.
2- Tissues samples: liver and kidney samples were collected and homogenized in phosphate buffer saline then

centrifuged at 3000 rp 15 min and homogenate preserved at -20 °C until used.

3- DNA sample: liver and kidney tissues were electrophoresed on 1.4 % agarose gel Tris-borate-EDRA buffer by the submarine gelelectrophoresis machine.

Biochemical analysis:

Determination of Alkaline phosphatase (ALP):

Serum ALP level was estimated using stanbio kit as described by **Tietz (1994).**

Determination of Serum amino transferases:

Serum ALT and AST levels were estimated using stanbio kit as described by **Young (1990).**

Determination of Serum urea:

Serum urea level was estimated using stanbio kit as described by **Tietz (1986).**

Determination of Serum creatinine:

Serum creatinine level was estimated using stanbio kit as described by **Tietz and Norbert (1995).**

Determination of Serum uric acid:

Serum uric acid level was estimated using stanbio kit as described by **Fossati *et al.* (1980).**

Determination of Serum total protein:

Serum total protein level was estimated using stanbio kit as described by **Orsonneau *et al.* (1989).**

Determination of Serum albumin:

Serum albumin level was estimated using stanbio kit as described by **Dumas and Biggs (1972).**

Determination of Serum globulin:

It was calculated as the difference between the total protein and albumin.

Determination of Albumin /Globulin ratio:

It was calculated from the albumin present in the serum in relation to the amount of globulin.

Determination of Serum glucose:

Glucose level was estimated using stanbio kit as described by **Young (1990).**

Determination of Serum cortisol:

Cortisol determined using RIA kit according to the method described by **Foster and Dunn (1974).** It was measured by using DSL-2100 ACTIVE®

Determination of LDH:

It was assessed by the method of **Vassault (1983).**

Serum lipogram analysis:

Determination of Cholesterol:

It was estimated as described by **Stein (1986).**

Determination of Triglycerides:

It was determined according to **Scheletter and Nussel (1975).**

Oxidative stress and antioxidant enzymes:

Tissue of liver and kidney were homogenized in a potter-Elvejhem glass/Teflon homogenizer and centrifuged then the supernatant was collected for determined the following parameters:

Determination of Lipid peroxidation (LPO):

It was estimated according to method described by **Ohkawa *et al.* (1979).** The method is based on the reaction of thiobarbituric acid (TBA) with malondialdehyde (MDA) in acidic medium at temperature of 95°C for 30 min to form thiobarbituric acid reactive produce the absorbance of the resultant pink product can be measured at 534 nm.

Determination of Reduction glutathione (GSH):

It was estimated according to the method described by **Beutler *et al.* (1963).** The method was based on the reduction of 5,5 ˉ dithiobis (2-nirtobenzoic acid) (DTNB) with glutathione (GSH) to produce a yellow compound. The reduced chromogen directly proportional to (GSH) concentration and its absorbance could be measured at 405 nm.

Determination of Catalase enzyme (CAT):

According to method described by **Aebi (1984)** the concentration of catalase enzyme was estimated by using Bio-Diagnostic kit. In this test, catalase reaction with a known quantity of H_2O_2. The reaction is stopped after exactly one minute with catalase inhibitor.

$$2H_2O_2 \xrightarrow{\text{Catalase}} 2H_2O + O_2$$

Determination of Superoxide dismutase (SOD):

The concentration of superoxide dismutase enzyme in liver and kidney tissues were measured calorimetrically according to the method described by **Nishikimi *et al.*, (1972).** This assay relied on the ability of the enzyme to inhibit the phenazine methosulphate-mediated reduction of nirtoblue tetrazolium dye.

Determination of Total anti-oxidant capacity (TAC):

TAC was determined using a commercial kit (Antioxidant assay kit, cat. No. CS0790; Sigma-Aldrich, USA). Absorbance was read at 405 nm using a Multiscan EX microplate reader (labsystems, Helsinki, Finland). Under conditions of this assay, the volume of plasma or diluted plasma was 10 µl and TAC values of samples were expressed as an equivalent of the mmol concentration of a (±)-6-hydroxy-2,5,7,8-tetramethylchromane-2-carboxylic acid (Trolox) solution. Trolox standard curve at the range of 0.00-0.42 mM was prepared for the assay according to the method described by **Koracevic *et al.* (2001).**

Agarose gel electrophoresis for DNA fragmentation:

The analysis of DNA fragmentation was carried out by agarose gel electrophoresis according to the method described by **Yokozawa and Dong (2001).** DNA samples of liver and kidneys tissues of catfish were electrophoresed on 1.4% agarose gel using Tris-borate- EDTA (TBE) buffer by the submarine gel electrophoresis machine and DNA was visualized and photographed by Ultra Vilot transilluminator.

Fragmented DNA was calculated from fragmented DNA+ O.D of intact absorbance reading at 600 nm using the formula.

DNA fragmentation % = O.D fragmented DNA/ (O.D fragmented DNA+ O.D of intact DNA) * 100.

Histopathological studies:

Tissue specimens from liver and kidneys were taken, fixed in 10 % neutral buffer formalin, dehydrated in alcohol, cleared in xylol and embedded in paraffin. the 4 um thick sections were prepared and stained with H&E **(Bancroft *et al.*, 1996 and Roberts, 2001).** Slides were examined for histological alterations.

Statistical Analysis:

The data were presented as means ± SE. Data were subjected to two-way ANOVA to evaluate effects of Ag-NPs toxicity and NS administration. followed by Tukey's-B post hoc test to compare between groups, $P<0.05$ was considered statistically significant. All the data analyses were done using **SPSS (2004)**.

RESULTS

96 hours half lethal concentration (LC_{50}) of silver nanoparticles in *Clarias gariepinus*:

In this study, the half lethal concentration (LC_{50}) of Ag-NPs is demonstrated in table (1)

Table (1): Half lethal concentration (LC_{50}) 96h of Ag-NPs for *C. gariepinus*.

Nano $AgNO_3$ (mg/L)	No. of fish for every concentration	No. of alive fish	No. of dead fish	a	b	a x b
0.00	8	8	0	0	0	0
41	8	7	1	41	0.5	20.5
82	8	5	3	41	2	82
164	8	3	5	82	4	328
328	8	1	7	164	6	984
656	8	0	8	328	7.5	2460

Where: n: Number of fish in each group.

a: Constant factor of difference between groups.

b: Mean value of dead fish between each two successive groups.

Fig. (1) Micrograph of Ag-NPs in the test TEM (100 nm &8000)

Measurement in nm Entries total. Mean = 12.4nm, Dev (rms) = 1.972nm

10.9nm, 12.7nm, 10.5nm, 15.5nm.

Fig. (2) Micrograph of Ag-NPs in the test TEM (100 nm &8000)

Measurement in nm Entries total. Mean = 6.899nm, Dev (rms) = 1.085nm.

7.37 nm, 7.65 nm, 8.28 nm, 4.69 nm, 6.44 nm, 7.41nm.

1
10.9 nm
15.5 nm
10.5 nm
12.7 nm
100 nm
TEM Mag = 80000x
S4 new.tif
Print Mag: 86600x @ 51 mm
100 nm
Direct Mag: 80000x
AMT Camera System

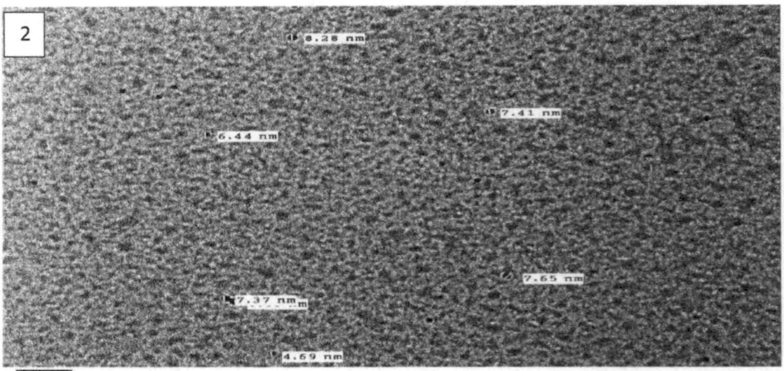
2
8.28 nm
7.41 nm
6.44 nm
7.65 nm
7.37 nm
4.69 nm
100 nm
TEM Mag = 100000x
s-14.tif
Print Mag: 108000x @ 51 mm
100 nm
HV=80.0kV
Direct Mag: 100000x
AMT Camera System

Clinical signs and postmortem findings:

In this study, *C. gariepinus* exposes to 50 mg/L of Ag-NPs for 30 days, showed abnormal skin pigmentation, moving towards the air pump, staying for a long time, a certain location also, irregular movement and loss of equilibrium whereas observed. Whereas, co-administration with 3% *Nigella sativa* in group (IV) reduced the abnormal behavior of catfish exposed to Ag-NPs.

Serum biochemical analysis:

Serum liver enzymes:

Serum Alkaline phosphatase (ALP):

The mean values of ALP in serum samples of *C. gariepinus* exposed to Ag-NPs 50 mg/L alone or co-administrated with NS 3% for 30 days are demonstrated in Table 2 and fig 3. The present results revealed that level of ALP was significantly ($P<0.05$) increased at the end of experiment in group (III). However, exposued *C. gariepinus* with 50 mg/L Ag-NPs and supplemented with 3% NS reduced the level of ALP to the normal level of control group.

Serum alanine amino transferase (ALT):

ALT level in serum samples of catfish exposed to 50 mg/L Ag-NPs alone or co-treated with 3% NS for 30 days are showed in Table 2 and fig 3. The data recorded that the level of ALT was significantly ($P<0.05$) increased at the end of experiment in group (III). Meanwhile, exposued catfish with 50 mg/L Ag-NPs and supplemented with 3% NS reduced the level of ALT to the normal level of control group.

Serum Aspartate amino transferase (AST):

Table 2 and fig 3 revealed that at the end of experiment AST level showed a significant increase ($P<0.05$) in exposed group (III). However, exposued *C. gariepinus* with 50 mg/L Ag-NPs and supplemented with NS 3% reduced the level of AST to the normal level of control group.

Table (2): Effect of 50 mg/L Ag-NPs on serum liver function in African catfish (*C. gariepinus)* and treated with 3% NS for 30 days.

Parameters / Groups	ALP (μ/l)	ALT (μ/l)	AST (μ/l)
Group I (Control)	$23.21^{b}\pm1.25$	$30.37^{b}\pm0.68$	$124.04^{a}\pm20.99$
Group II (3% NS)	$28.44^{b}\pm5.64$	$28.57^{b}\pm0.70$	$103.85^{a}\pm6.11$
Group III (Ag/NPs)	$48.22^{a}\pm2.89$	$76.67^{a}\pm10.41$	$155.07^{a}\pm3.71$
Group IV (3%NS+ Ag/NPs)	$24.31^{b}\pm3.41$	$38.50^{b}\pm3.75$	$126.44^{a}\pm21.88$

Data were founded as mean ± SE (n=10) values in the same column for every group are ($P<0.05$).

a= significant change from the control value

b= significant change from the control NS

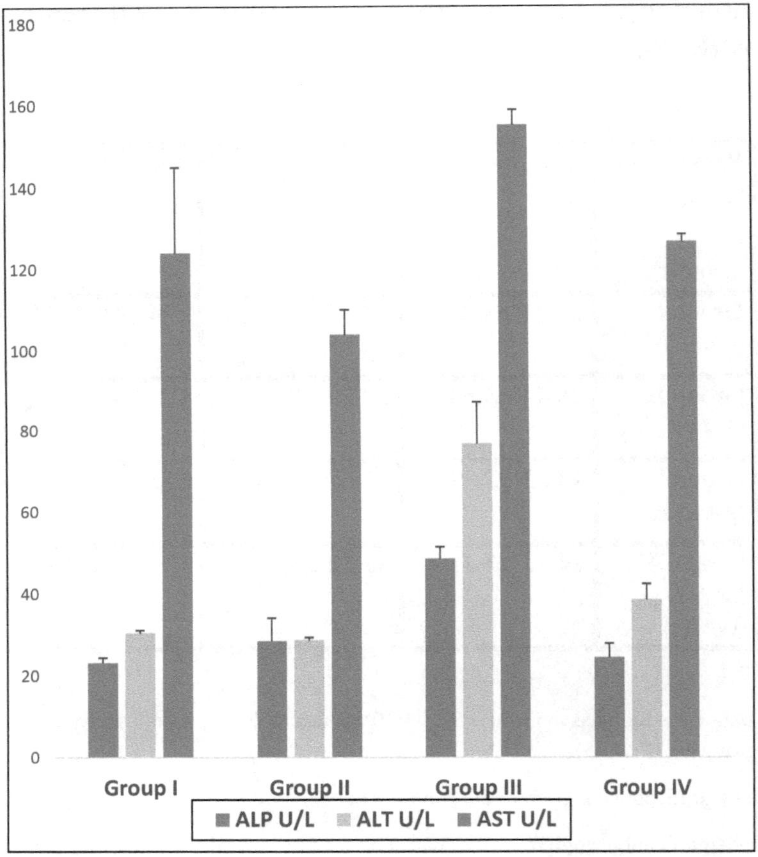

Fig. (3): Histogram showing the variations in ALP (U/L), ALT (U/L), and AST (U/L) levels of *C. gariepinus* exposed to 50 mg/L Ag-NPs and treated with 3% NS for 30 days.

Serum kidney function:

Serum urea:

The present results show that after 30 days of exposed to 50 mg/L Ag-NPs, there was a significant increase in urea level in group (III). However, exposed catfish with Ag-NPs 50 mg/L and treated with 3% NS reduced the level of urea to the normal level of control group (Table 3 and fig 4).

Serum creatinine:

The mean values of creatinine in serum samples of *C. gariepinus* exposed to 50 mg/L Ag-NPs alone or co-treated with 3% NS after 30 days are demonstrated in Table 3 and fig 5. The results showed that the level of creatinine was markedly ($P<0.05$) increased in group (III). Meanwhile, exposed *C. gariepinus* with 50 mg/L Ag-NPs and supplemented with 3% NS reduced the level of creatinine to the normal level of control group.

Serum uric acid:

It can be see from Table 3 and fig 5 that the values of uric acid in serum samples of catfish exposed to 50 mg/L Ag-NPs alone or co-treated with 3% NS show that the level of uric acid was markedly ($P<0.05$) increase in group (III) as compared to the control group. Whenever, exposed catfish with Ag-NPs 50 mg/L and treatment with 3% NS decreased the level of uric acid to the normal level of control group.

Table (3): Effect of 50 mg/L Ag-NPs on serum kidney function of African catfish (*C. gariepinus*) and treated with 3% NS for 30 days.

Parameters / Groups	urea (mg/dl)	Creatinine (mg/dl)	uric acid (mg/dl)
Group I (Control)	18.50^{b}±2.36	0.30^{a}±0.04	1.09^{b}±0.14
Group II (3% NS)	19.40^{ab}±0.56	0.32^{a}±0.02	1.52^{ab}±0.27
Group III (Ag/NPs)	23.20^{a}±1.10	0.41^{a}±0.01	2.11^{a}±0.22
Group IV (3%NS+ Ag/NPs)	20.27^{ab}±0.32	0.36^{a}±0.04	1.33^{b}±0.21

Data are responded as mean ± SE (n=10) values in the same column for every group are ($P<0.05$).

a= significant change from the control value

b= significant change from the control NS

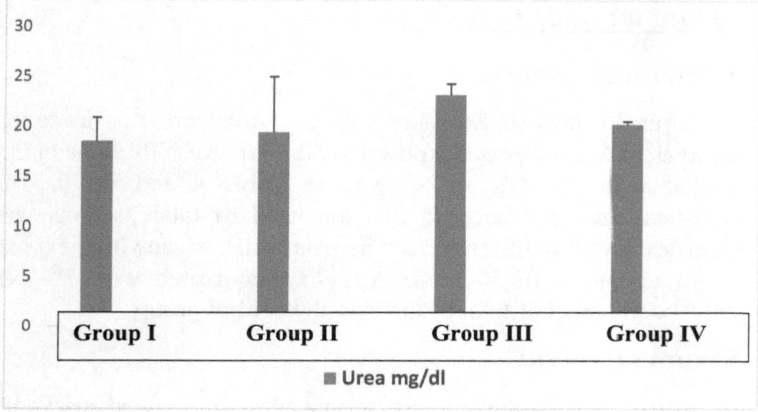

Fig. (4) Histogram showing the variation in urea (mg/dl) level of *C. gariepinus* exposed to 50 mg/L Ag-NPs and treated with 3% NS for 30 days.

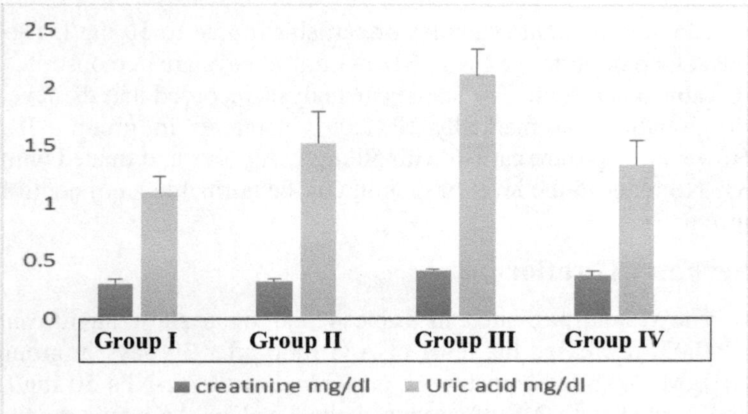

Fig. (5): Histogram showing the variations in Creatinine (mg/dl), and uric acid (mg/dl) levels of *C. gariepinus* exposed to 50 mg/L Ag-NPs and treated with 3% NS for 30 days.

Serum immune function:

Serum total proteins:

After 30 days of exposure values of total proteins in serum samples of *C. gariepinus* exposed to 50 mg/L Ag-NPs alone or co-treated with 3% NS are showed in Table 4 and fig 6. The statistical analysis recorded that the level of total proteins was significantly ($P<0.05$) increased in group (III). Meanwhile, exposd *C. gariepinus* with 50 mg/L Ag-NPs and treated with 3% NS reduced the level of total proteins to the control group.

Serum albumin:

From (Table 4 and Fig 6), serum albumin showed markedly ($P<0.05$) increased after 30 days in group (III). However, exposed catfish with 50 mg/L Ag-NPs and supplemented with 3% NS reduced the level of albumin to the normal level of control group.

Serum globulin:

Globulin in serum samples of catfish exposed to 50 mg/L Ag-NPs alone or co-treated with 3% NS for 30 days are demonstrated in Table 4 and fig 6. The statistical analysis recorded that the level of globulin was markedly ($P<0.05$) increased in group (III). However, exposure catfish with 50 mg/L Ag-NPs and treated with 3% NS reduced the level of globulin to the normal level of control group.

Serum A/G ratio:

The results presented in Table 4 and fig 6 show significant ($P<0.05$) decreased the level of A/G ratio after 30 days in group (III). Meanwhile, exposure *C. gariepinus* with Ag-NPs 50 mg/L and treated with NS 3% reduced the level of A/G ratio to the control group.

Table (4): Effect of 50 mg/L Ag-NPs on serum Total protein, albumin, globulin and A/G ratio of African catfish (*C. gariepinus*) and treated with 3% NS for 30 days.

Parameters / Groups	total protein (g/dl)	Albumin (g/dl)	Globulin (g/l)	A/G ratio
Group I (Control)	$7.11^{b}\pm0.40$	$2.70^{b}\pm0.10$	$4.41^{b}\pm0.09$	$0.61^{a}\pm0.09$
Group II (3% NS)	$8.12^{ab}\pm0.57$	$3.03^{ab}\pm0.12$	$5.09^{b}\pm0.27$	$0.59^{a}\pm0.23$
Group III (Ag/NPs)	$9.37^{a}\pm0.25$	$3.35^{a}\pm0.14$	$6.02^{ab}\pm0.06$	$0.55^{a}\pm0.06$
Group IV (3%NS+ Ag/NPs)	$8.77^{a}\pm0.46$	$3.27^{a}\pm0.08$	$5.5^{a}\pm0.26$	$0.59^{a}\pm0.11$

Data are responded as mean ± SE (n=10) values in the same column for every group are ($P<0.05$).

a= significant change from the control value

b= significant change from the control NS

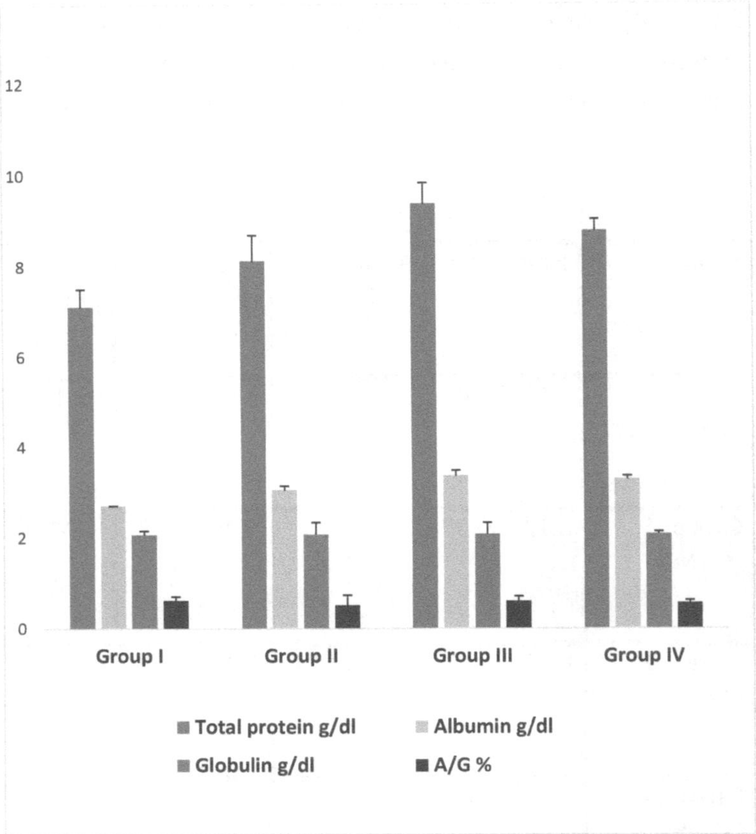

Fig. (6) Histogram showing the variations in Total protein (g/dl), Albumin (g/dl), Globulin (g/dl), and A/G ratio levels of *C. gariepinus* exposed to 50 mg/L Ag-NPs and treated with 3% NS for 30 days.

Serum stress indicators:

Serum glucose:

Glucose in serum samples of catfish exposed to 50 mg/L (Ag-NPs) alone or co-treated with 3% NS for 30 days are demonstrated in Table 5 and fig 7. The results revealed that the level of glucose was significantly ($P<0.05$) increased in group (III). However, exposure catfish with 50 mg/L (Ag-NPs) and treated with 3% NS reduced the level of glucose to the normal level of control group.

Serum cortisol:

It can be seen from Table 5 and fig 7 that cortisol level in serum samples was significantly ($P<0.05$) increased after 30 days in group (III). Meanwhile, exposure *C. gariepinus* with 50 mg/L Ag-NPs and treated with 3% NS reduced the level of cortisol to the normal level of control group.

Table (5): Effect of 50 mg/L Ag-NPs on serum Glucose and cortisol of African catfish (*C. gariepinus*) and treated with 3% NS for 30 days.

Parameters / Groups	Glucose (mg/dl)	Cortisol (µg/dl)
Group I (Control)	67.33^{b}±5.61	4.77^{b}±0.66
Group II (3% NS)	79.00^{b}±6.66	18.07^{a}±5.19
Group III (Ag/NPs)	125.00^{a}±8.14	27.22^{a}±4.00
Group IV (3%NS+ Ag/NPs)	113.00^{a}±7.51	18.09^{a}±0.75

Data are responded as mean ± SE (n=10) values in the same column for every group are (P<0.05).

a= significant change from the control value

b= significant change from the control NS

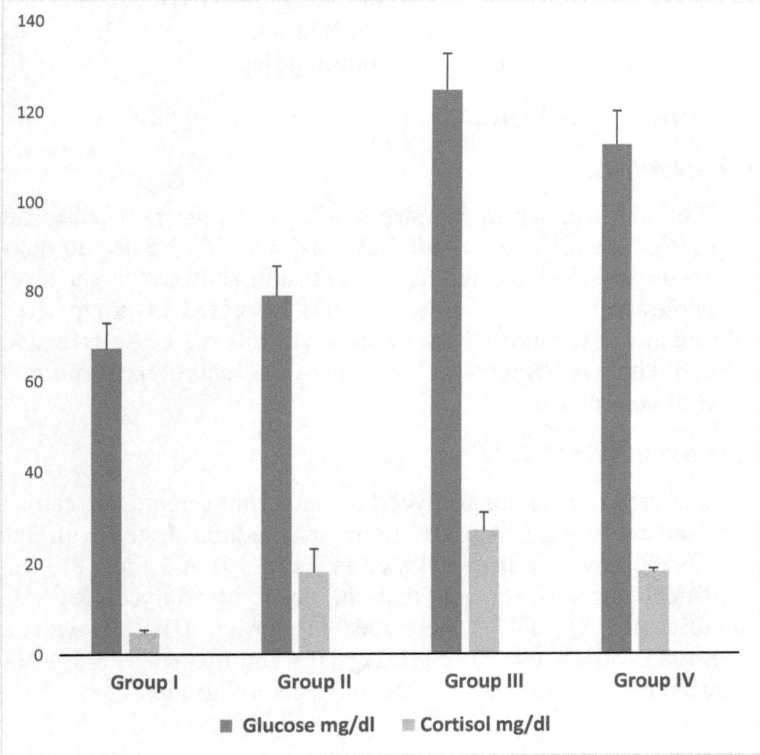

Fig. (7) Histogram showing variations in glucose (mg/dl), and cortisol hormone (mg/dl) levels of *C. gariepinus* exposed to 50 mg/l Ag-NPs and treated with 3% NS for 30 days.

LDH enzyme:

The results presented in Table 6 and fig 8 show a significantly ($P<0.05$) increased LDH enzyme in group (III). Meanwhile, exposure catfish with 50 mg/L Ag-NPs and treated with 3% NS reduced the level of LDH to the control group.

Serum Total lipids:

Cholesterol:

Cholesterol in serum samples of *C. gariepinus* exposed to 50 mg/L Ag-NPs alone or co-administrated with 3% NS for 30 days are recorded in Table 6 and fig 8. The results showed that the level of cholesterol was markedly ($P<0.05$) increased in group (III). Meanwhile, exposure *C. gariepinus* with 50 mg/L Ag-NPs and treated with 3% NS reduced the level of cholesterol to the normal level of control group.

Triglycerides:

The mean values of triglycerides in serum samples of catfish exposed to 50 mg/L Ag-NPs alone or co-administrated with 3% NS for 30 days are demonstrated in Table (6) and Fig. (8). The statistical analysis showed that the level of triglycerides was significantly ($P<0.05$) increased in group (III). However, exposued catfish with 50 mg/L Ag-NPs and treated with 3% NS reduced the level of triglycerides to the control group.

Table (6): Effect of 50 mg/L Ag-NPs on LDH enzyme, cholesterol and triglycerides of African catfish (*C. gariepinus*) and treated with 3% NS for 30 days.

Parameters / Groups	LDH (μ/l)	Cholesterol (mg/dl)	Triglycerides (mg/l)
Group I (Control)	$651.67^{b}\pm50.85$	$334.67^{b}\pm45.67$	$148.84^{a}\pm16.74$
Group II (3% NS)	$952.00^{b}\pm332.82$	$291.67^{b}\pm42.06$	$141.33^{a}\pm12.41$
Group III (Ag/NPs)	$1609.70^{a}\pm175.43$	$585.33^{a}\pm104.99$	$202.67^{a}\pm30.12$
Group IV (3%NS+ Ag/NPs)	$634.33^{b}\pm35.57$	$276.00^{b}\pm12.22$	$177.00^{a}\pm13.87$

Data are responded as mean ± SE (n=10) values in the same column for every group are (P<0.05).

a= significant change from the control value

b= significant change from the control NS

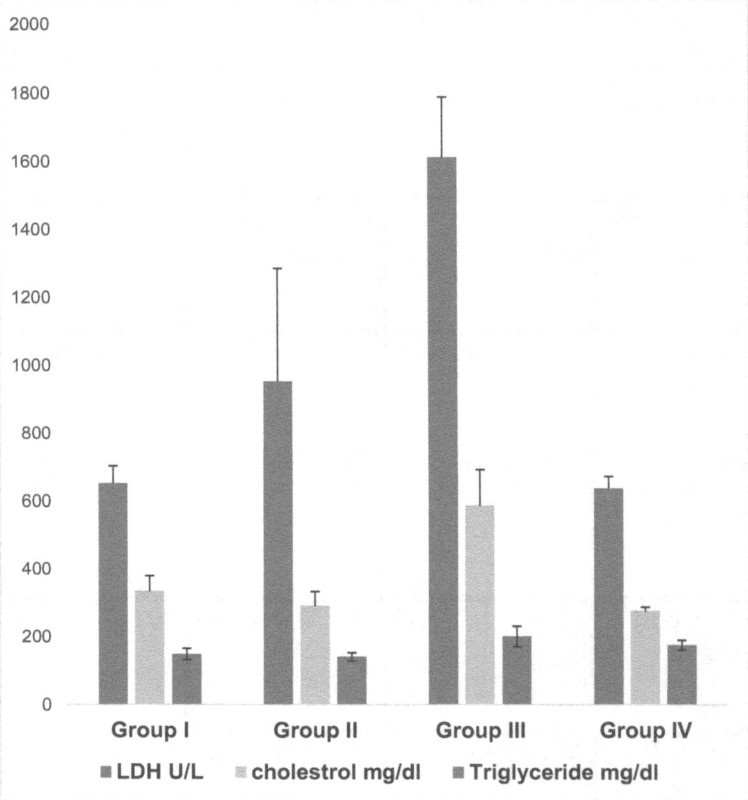

Fig. (8): Histogram showing variations in LDH (U/L), cholesterol (mg/dl), and Triglycerides (mg/dl) levels of *C. gariepinus* exposed to 50 mg/l Ag-NPs and treated with 3% NS for 30 days.

Oxidative stress and antioxidant enzymes:

Lipid peroxidation (LPO):

The results presented in Tables (7 & 8) and figs (9&11) show that significantly increased LPO level ($P<0.05$) in the liver and kidney tissues of catfish in group (III) after exposed to 50 mg/L Ag-NPs for 30 days as compared to the tissues of hepato-renal control. However, NS 3% NS treated improved LPO to nearly normal levels.

Reduced glutathione (GSH):

The mean values of hepato-renal GSH in serum samples of *C. gariepinus* exposed to 50 mg/L Ag-NPs alone or co-administrated with 3% NS for 30 days are demonstrated in Tables (7 & 8) and figs (9& 11). Levels of GSH were markedly increased ($P<0.05$) in the liver and kidney tissues in group (III) as compared to the tissues of liver and kidney of control. In order that, 3% NS administration improved GSH to nearly normal levels.

Catalase enzyme (CAT):

CAT level in serum samples of catfish exposed to Ag-NPs 50 mg/L alone or co-treated with 3% NS for 30 days are recorded in Tables (7&8) and Figs (9&11). Levels of CAT were increased ($P<0.05$) in the liver and kidney tissues in group (III) as compared to the tissues of hepato-renal control group. Whenever, 3% NS treated CAT to normal levels.

Superoxide dismutase (SOD):

The mean values of hepato-renal SOD in samples of *C. gariepinus* exposed to 50 mg/L Ag-NPs alone or co-treated with 3% NS for 30 days are showed in Tables (7 & 8) and Figs. (10&

12). Levels of SOD were markedly increased ($P<0.05$) in the liver and kidney tissues in group (III) as compared to control group. In order that, 3% NS administration improved SOD to nearly normal levels.

Total Anti-Oxidant Capacity (TAC):

Tables (7 & 8) and Figs (10& 12) significantly increased TAC level ($P<0.05$) in the liver and kidney tissues of catfish after 30 days of exposure of 50 mg/L Ag-NPs in group (III) as compared to the tissues of hepato-renal control group. However, 3% NS administration improved TAC to nearly normal levels.

Table (7): Effect of 50 mg/L Ag-NPs on oxidative stress and antioxidant enzymes in liver of African catfish (*C. gariepinus*) and treated with 3% NS for 30 days.

parametrs / Groups	LPO (nmol/g)	GSH (nmol/mg)	CAT (Ug/mg)	SOD (Ug/mg)	TAC (Umol/mg)
Group I (Control)	$16.26^{d}\pm0.72$	$15.48^{c}\pm0.82$	$7.73^{c}\pm0.63$	$1.34^{c}\pm0.19$	$1.22^{b}\pm0.07$
Group II (3% NS)	$42.54^{c}\pm1.98$	$17.96^{ab}\pm1.06$	$17.10^{b}\pm0.40$	$4.73^{a}\pm0.23$	$2.80^{a}\pm0.05$
Group III (Ag/NPs)	$63.31^{b}\pm1.09$	$20.54^{a}\pm1.03$	$37.50^{a}\pm2.85$	$5.02^{a}\pm0.45$	$3.22^{a}\pm0.38$
Group IV (3%NS+ Ag/NPs)	86.67a±2.13	$18.77^{a}\pm0.62$	$22.40^{b}\pm1.42$	$3.08^{b}\pm0.09$	$2.48^{a}\pm0.24$

Data are responded as mean ± SE (n=10) values in the same column for every group are ($P<0.05$).

a= significant change from the control value

b= significant change from the control NS

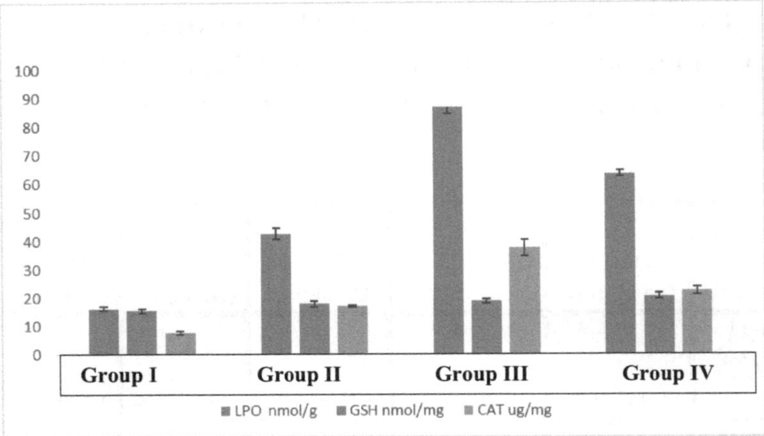

Fig. (9) Histogram showing the variations in liver LPO (nmol/g), GSH (nmol/mg), and CAT (Ug/mg) levels of *C. gariepinus* exposed to 50 Ag-NPs and treated with 3% NS for 30 days.

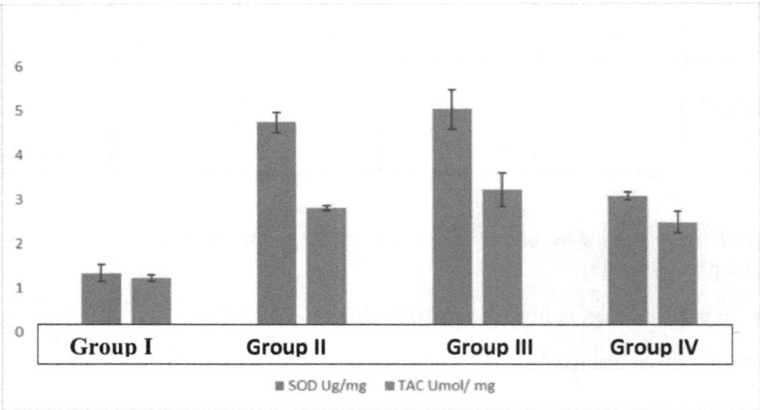

Fig. (10) Histogram showing the variations in liver SOD (Ug/mg), and TAC (Umol/mg) levels of *C. gariepinus* exposed 50 Ag-NPs and treated with 3% NS 30 days.

Table (8): Effect of 50 mg/L Ag-NPs on oxidative stress and antioxidant enzymes in kidney of African catfish (*C. gariepinus*) and treated with 3% NS for 30 days.

Parameter / Groups	LPO (nmol/g)	GSH (nmol/mg)	CAT (Ug/mg)	SOD (Ug/mg)	TAC (Umol/mg)
Group I (Control)	$59.25^{c}\pm0.47$	$1.88^{c}\pm0.17$	$19.51^{d}\pm0.44$	$2.82^{c}\pm0.17$	$28.69^{d}\pm0.70$
Group II (3% NS)	$68.67^{b}\pm0.84$	$7.84^{a}\pm0.93$	$26.18^{c}\pm1.59$	$6.94^{b}\pm0.38$	$38.93^{c}\pm0.52$
Group III (Ag/NPs)	$87.64^{a}\pm2.46$	$8.10^{a}\pm0.18$	$46.16^{a}\pm1.13$	$8.51^{a}\pm0.23$	$66.62^{a}\pm4.57$
Group IV (3%NS+ Ag/NPs)	$60.84^{c}\pm0.53$	$5.37^{b}\pm0.25$	$35.30^{b}\pm2.28$	$3.01^{c}\pm0.15$	$47.06^{b}\pm1.10$

Data are responded as mean ± SE (n=10) values in the same column for every group are ($P<0.05$).

a= significant change from the control value

b= significant change from the control NS

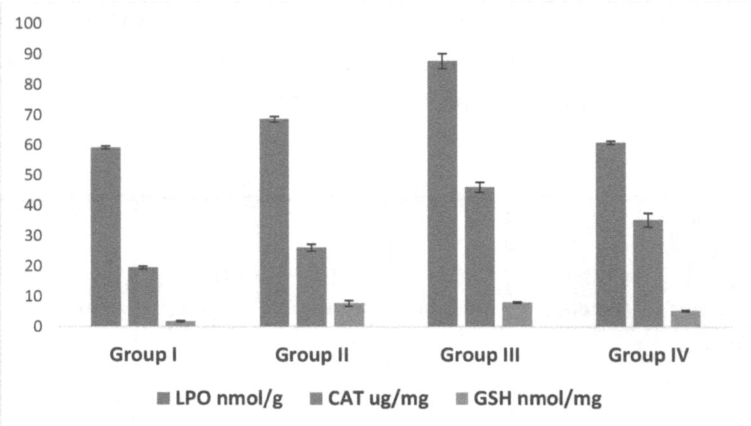

Fig. (11) Histogram showing the variations in kidney LPO (nmol/g), GSH (nmol/mg), and CAT (Ug/mg) levels of *C. gariepinus* exposed to 50 Ag-NPs and treated with 3% NS for 30 days.

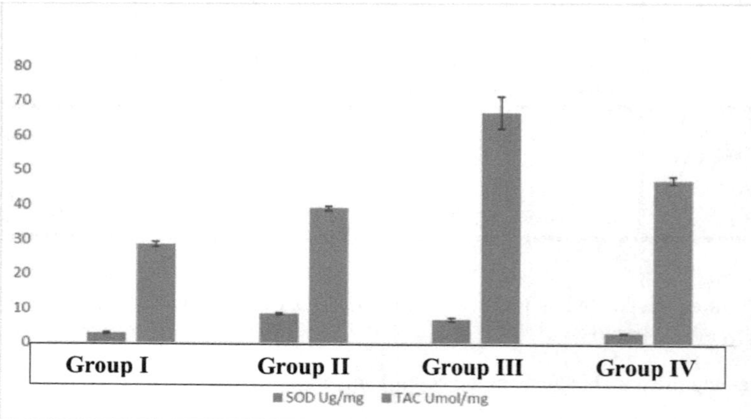

Fig. (12) Histogram showing the variations in kidney SOD (Ug/mg), and TAC (Umol/mg) levels of *C. gariepinus* exposed to 50 mg/L Ag-NPs and treated with 3% NS for 30 days.

DNA fragmentation analysis:

DNA fragmentation of liver samples of group (III) exposed to 50mg/L Ag-NPs; well: 6 was significantly increased compared to the control groups; lane: 2 and lane: 4 treated with 3% NS. However, lane: 8 of group (IV) exposed to 50mg/L Ag-NPs and fed on 3 % NS exhibited a marked reduction in hepatic DNA fragmentation. On the other hand, DNA fragmentation of kidney samples of group (III); lane: 7, respectively was increased markedly compared to the control groups; lane: 3 and lane: 5 was treated with 3% NS. Conversely, lane: 9 of group (IV) fed on 3 % NS displayed a significant reduction in renal DNA fragmentation. Fig. (13).

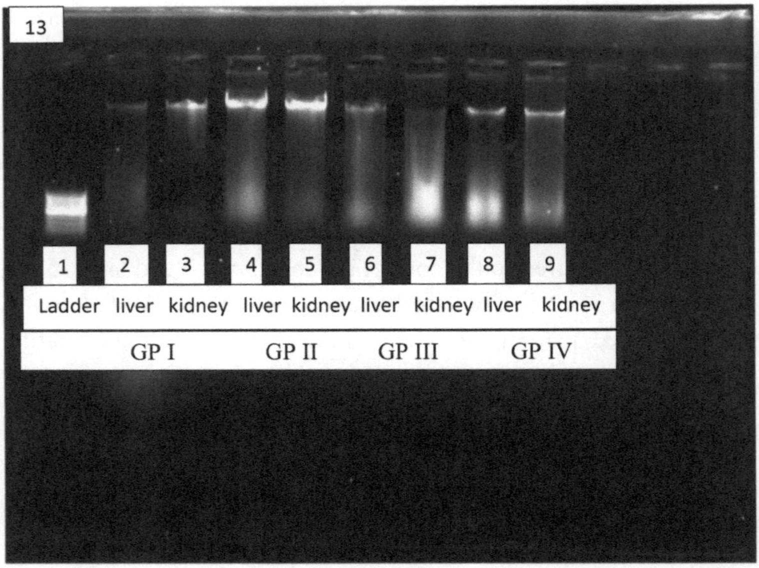

Fig. 13: Agarose gel electrophoresis showing DNA fragmentation pattern in liver and kidney tissues of catfish. lane1: Ladder; lane: 2, 3 for liver and kidney, respectively of group (I) fed on basal diet. lane: 4, 5 for liver and kidney, respectively of group (II) fed on 3% NS. lane: 6, 7 for liver and kidney, respectively of group (III) exposed to 50 mg/L Ag-NPs. lane: 8, 9 for liver and kidney, respectively of group (IV) exposed to 50mg/L Ag-NPs and fed on 3 % NS for 30 days.

Histopathological change:

The histopathology slides showed changes in the African catfish (*C. gariepinus*) liver and kidney tissues.

In liver:

As showed from Group (I) which is the control group (figs.14). And group (II) which co-treated with 3% NS represented a normal structure of hepatic tissues (Figs.15). Whereas, Liver tissues of *C. gariepinus* exposed to 50 mg/L Ag-NPs in group (III) showed hepatocytes (H) with hydropic degeneration and showed inflammatory cells (IF), melanomacrophage (Mm), beside central vein (CV), degeneration of the patocytes (D), cytoplasmic vaculation (CPV), rupture of the wall of the central vein (RCV), and melanomacrophage (Figs.16 & 17).

However, *C. gariepinus* of group (IV) exposed to Ag-NPs 50 mg/L and fed on NS 3% for 30 days exhibited in liver tissues (Figs. 18 & 19), hepatocytes (H) with mild hydropic degeneration. A single portal tract seen showing moderate inflammatory cellular infiltrate, melanomacrophage (Mm), moderate fibrosis (IF), and beside central vein (CV).

Fig. (14) Section in liver tissue of group (I) which is normal African catfish (*C. gariepinus*) showing general structure. Hepatocytes (H) and blood sinusoids (BS). (H&E, 400x).

Fig. (15) Section in liver tissue of African catfish (*C. gariepinus*) group (II) which is fed on 3% NS showing normal hepatocytes (H) and central vein (CV). (H&E, 200x).

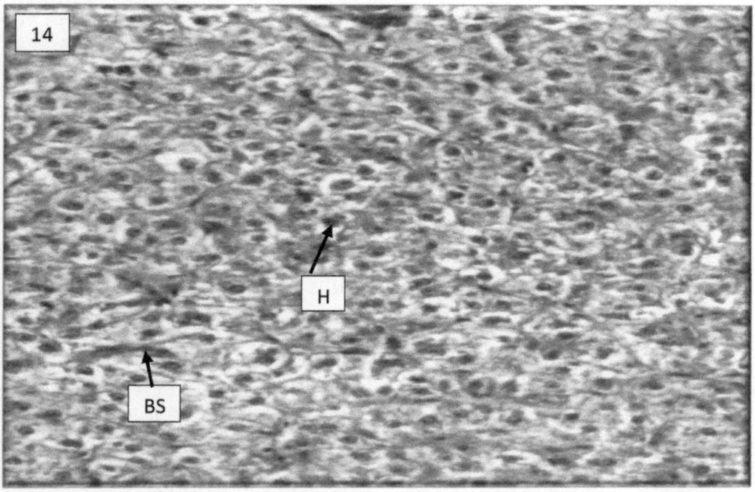
14
H
BS

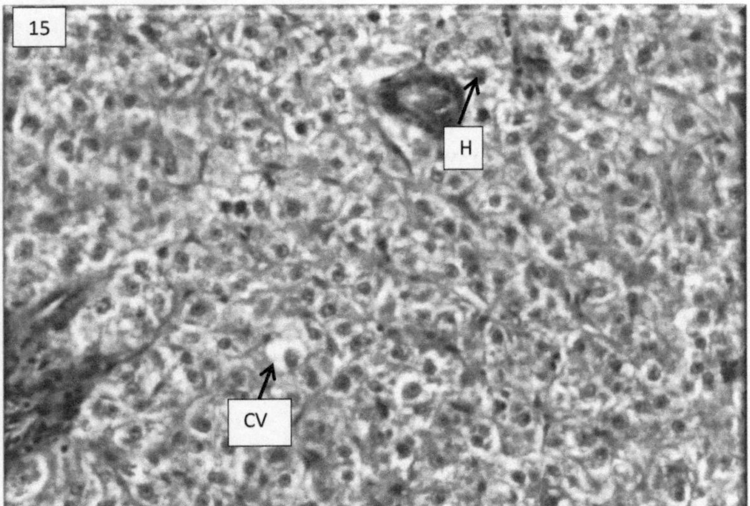
15
H
CV

Fig. (16) Section in liver tissue of African catfish (*C. gariepinus*) group (III) which is exposed to 50 mg/L Ag-NPs showing hepatocytes (H) with hydropic degeneration, inflammatory cells (IF), beside central vein (CV), melanomacrophage (Mm) and cytoplasmic vaculation (CPV), (H&E, 400x).

Fig. (17) Section in liver tissue of African catfish *(C. gariepinus*) group (III) which is exposed to 50 mg/L Ag-NPs showing hepatocytes (H) with marked hydropic degeneration, inflammatory cells (IF), degeneration of the patocytes (D), rupture of the wall of the central vein (RCV), melanomacrophage and mild fibrosis. (H&E, 400x).

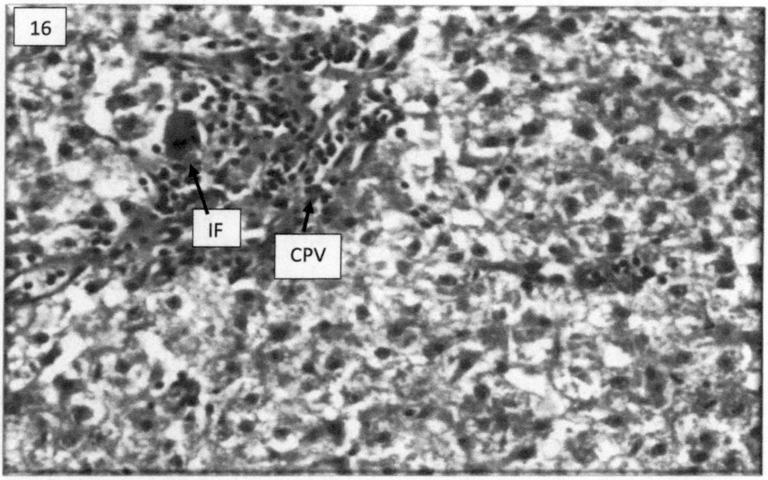
16
IF
CPV

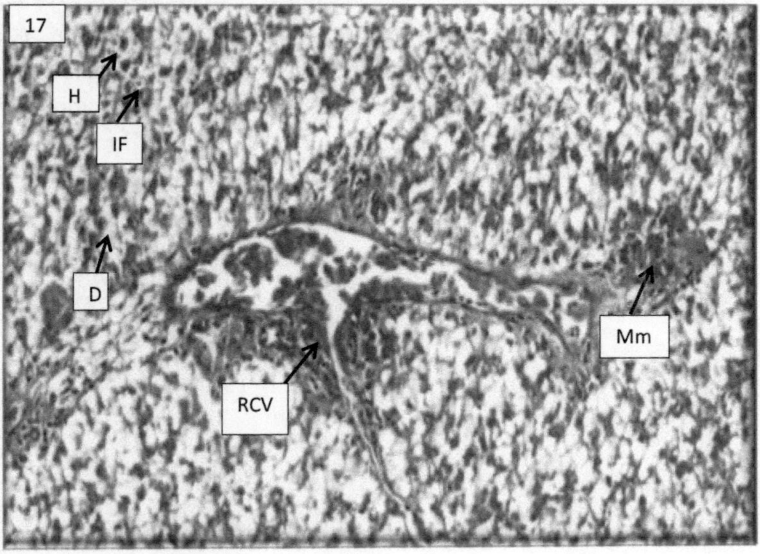
17
H
IF
D
RCV
Mm

Fig. (18) Section in liver tissue of group (IV) African catfish (*C. gariepinus*) exposed to 50 mg/L Ag-NPs and fed on 3% NS showing hepatocytes (H) with mild hydropic degeneration. A single portal tract seen showing moderate inflammatory cellular infiltrate, moderate fibrosis (IF), melanomacrophage (Mm) and central vein (CV). (H&E, 200x).

Fig. (19) Section in liver tissue of African catfish *(C. gariepinus)* group (IV) which exposed to 50 mg/L Ag-NPs and treated with 3% NS showing hepatocytes (H) with minimal hydropic degeneration. Inflammatory cellular filtrate (IF), melanomacrophage (Mm) and central vein (CV). (H&E, 400x).

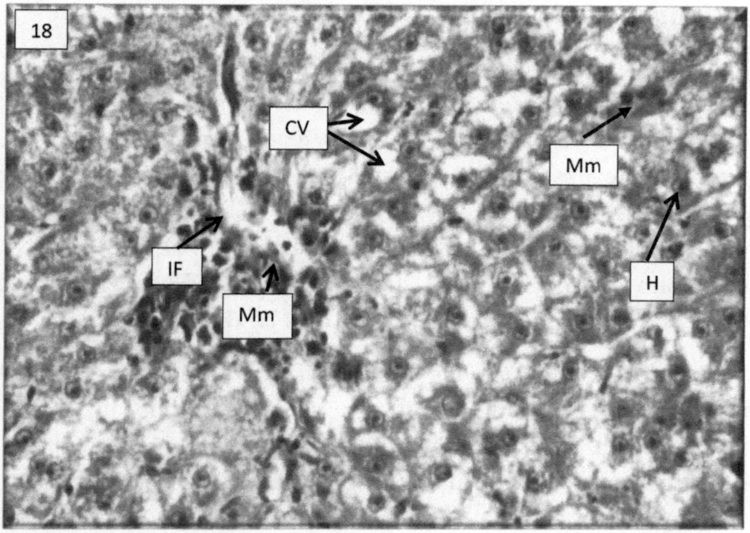
18
CV
Mm
IF
Mm
H

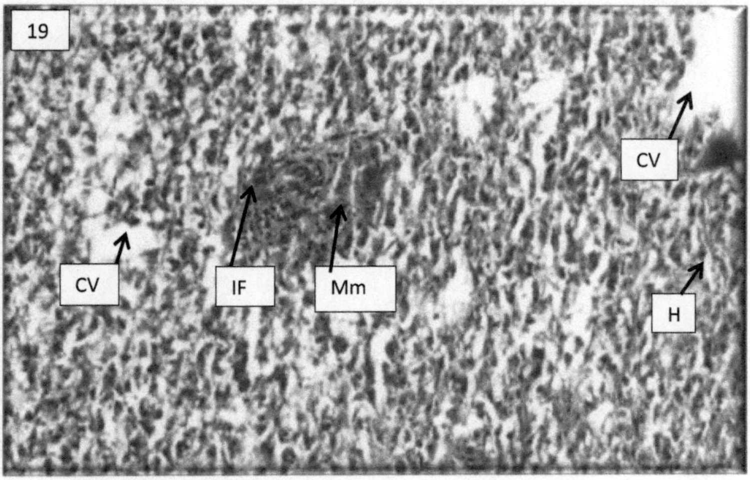
19
CV
CV
IF
Mm
H

In kidney:

As showed from Group (I) which is the control group (figs. 20). And group (II) which fed on 3% NS representated a normal structure of renal tissues (Figs. 21). Whereas, kidney tissues of *C. gariepinus* exposed to 50 mg/L Ag-NPs in group (III) showing obsolescent glomeruli (G) with mild inflammatory cellular infiltrate, inflammatory basement membrane (BM), necrosis (N), melanomacrophages (Mm), slight dilation of Bowmen's space (BS), and haemopoietic tissue (HT) (Figs 22 & 23).

However, *C. gariepinus* of group (IV) exposed to Ag-NPs 50 mg/L and treated with NS 3% for 30 days exhibited in kidney tissues (Figs. 24 & 25) showing patent glomeruli (G) with no inflammatory cellular infiltrate. slight dilation of Bowman's space (BS), The renal tubules lined by regular epithelium, melanomacrophages (Mm), and haemopoietic tissue.

Fig. (20) Section in kidney tissue of group (I) which is normal African catfish *(C. gariepinus)* showing patent glomeruli (G) with no inflammatory cellular infiltrate. The renal tubules (RT) lined by regular epithelium, normal basement membrane (BM) of the renal tubules, renal corpuscle (RC) and haemopoietic tissue (HT). (H&E, 200 x).

Fig. (21) Section in kidney tissue of African catfish *(C. gariepinus)* group (II) which is fed on 3% NS showing patent glomeruli (G) with no inflammatory cellular infiltrate. The tubules lined by regular epithelium, normal basement membrane (BM) of renal corpuscle (RC) and normal haemopoietic tissue (HT). (H&E, 200x).

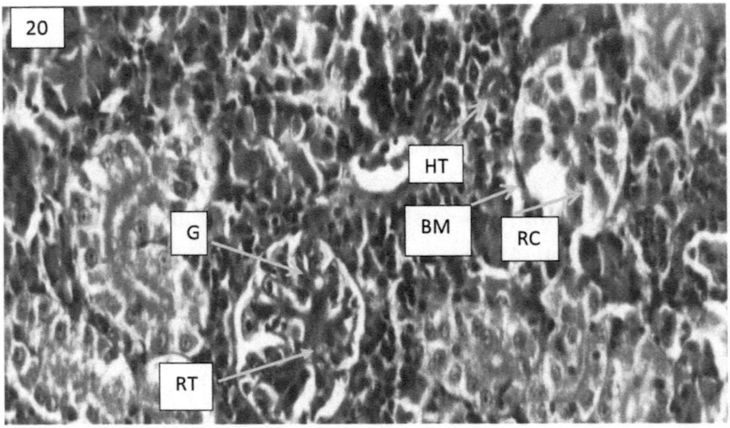
20
HT
BM
RC
G
RT

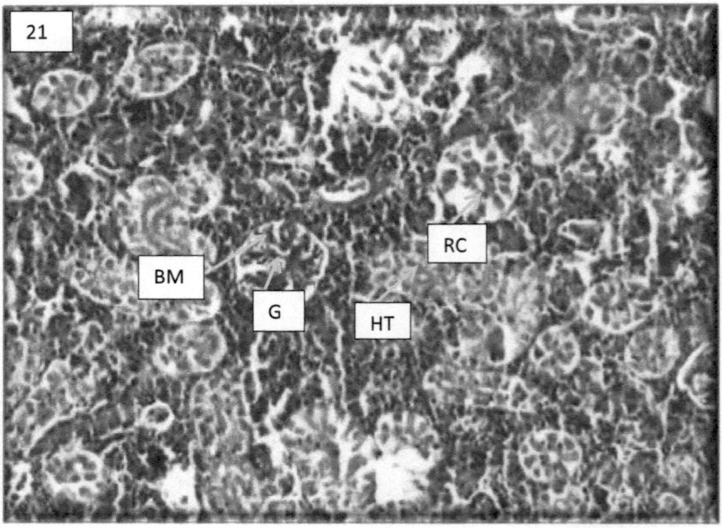
21
RC
BM
G
HT

Fig. (22) Section in kidney tissue of African catfish *(C. gariepinus)* group (III) which exposed to 50 mg/L Ag-NPs showing obsolescent glomeruli (G) with mild inflammatory cellular infiltrate and inflammatory basement membrane (BM), slight dilation of Bowmen's space (B), necrosis (N), melanomacrophages (Mm) and haemopoietic tissue (HT) (H&E, 200x).

Fig. (23) Section in kidney tissue of African catfish *(C. gariepinus)* group (III) which exposed to 50 mg/L Ag-NPs showing obsolescent glomeruli (G) with moderate inflammatory cellular infiltrate, necrosis (N), irregular renal tubules (RT) and melanomacrophages (Mm). (H&E, 200 x).

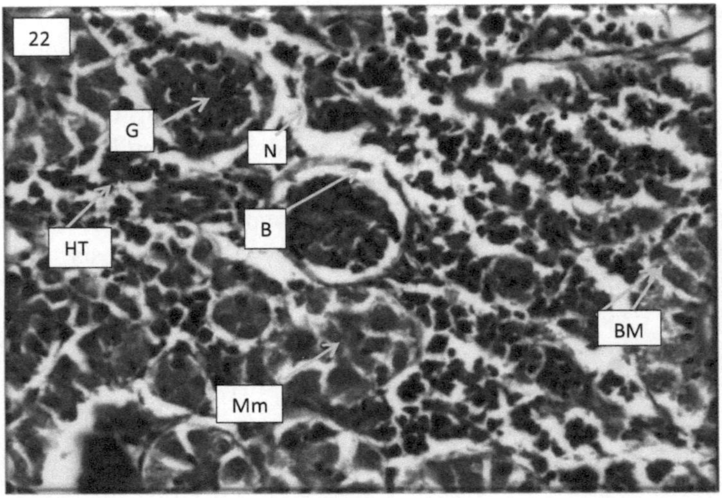
22
G
N
B
HT
BM
Mm

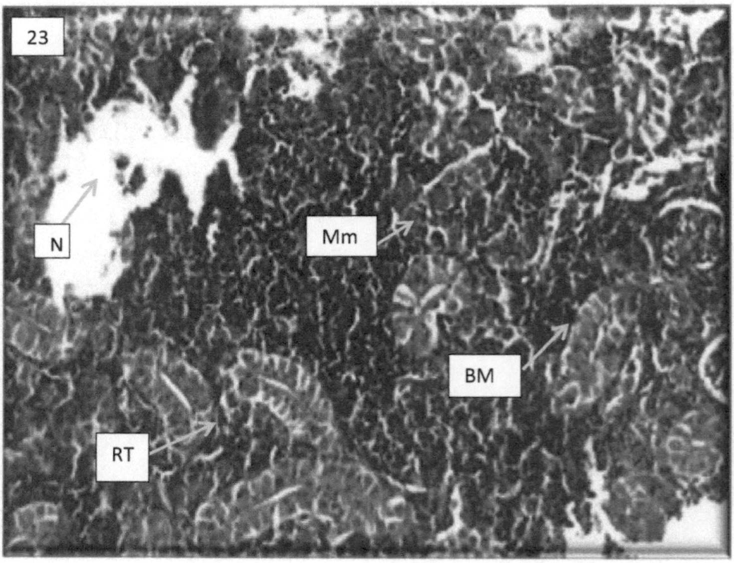
23
N
Mm
BM
RT

Fig. (24) Section in kidney tissue of African catfish *(C. gariepinus)* group (IV) which exposed to 50 mg/L Ag-NPs and fed on 3% NS showing patent glomeruli (G) with no inflammatory cellular infiltrate. The renal tubules lined by regular epithelium, slight dilaton of Bowman's space (B), melanomacrophages (Mm) and haemopoietic tissue (H&E, 400 x).

Fig. (25) Section in kidney tissue of African catfish *(C. gariepinus)* group (IV) which exposed to 50 mg/L Ag-NPs and fed on 3% NS showing patent glomeruli (G) with no inflammatory cellular infiltrate. The renal tubules lined by regular epithelium, melanomacrophages (Mm) and haemopoietic tissue (HT). (H&E, 400 x).

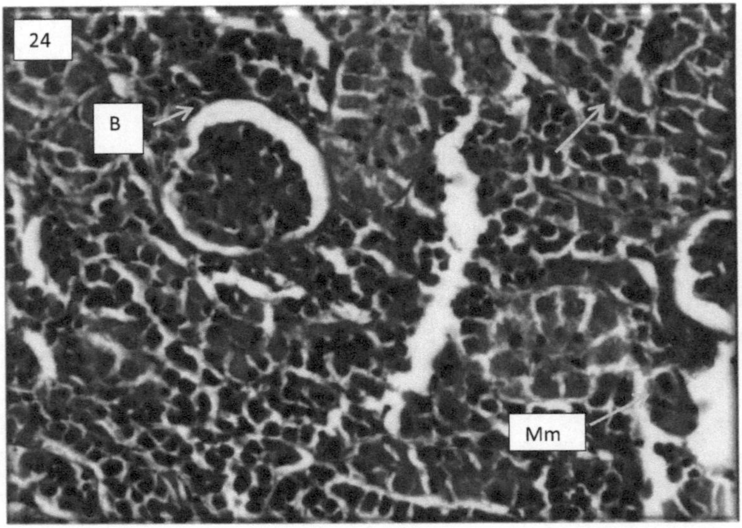
24
B
Mm

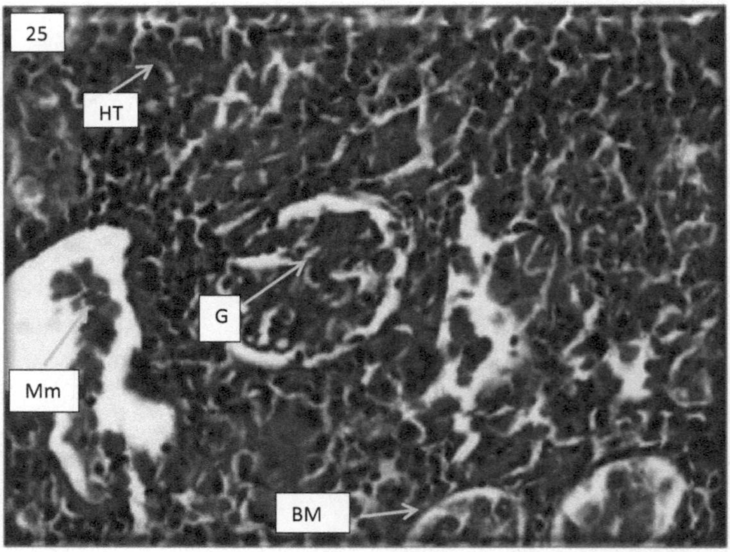
25
HT
G
Mm
BM

DISCUSSION

To assess the toxic effect of Ag-NPs, understand the impact of more conditions (exposuré conditions, aquatic organism type and particle type) and mechanisms of toxicity **(Vignesh *et al.,* 2013)**. Silver nanoparticles has been consolidated into consumer products including materials, air and water filters, medical equipment, contraceptives, beauty care products, youngsters' toys, and residential washing machines (**Stensberg *et al.,* 2011**). The usage of consumer products containing silver nanoparticles (Ag-NPs) indicates that Ag-NPs may be released into the aquatic ecosystem (**Kaegi *et al.,* 2010**). The consequent emission of Ag-NPs into the aquatic environment which makes it necessary to assess the environmental and health hazards that these compounds could exert. Ag-NPs enter organism's body through oral absorption from food containing silver or damaged skin **(Drake and Hazelwood, 2005).**

All changes occur in catfish by Ag-NPs can improve by *Nigella sativa* seeds as in present study. Black seed is herbal medicinal plants has cytoprotective effects and is one of the safest and widely used medicinal product, it has a cure for every disease expect death **(Goreja, 2003). Abd-Elkareem *et al.* (2020)** showed the role of NS on African catfish (*C. gariepinus*) against toxicity and it is role in modulation, preventation and treatment of the effect of environmental toxins in aquatic animal.

Half lethal concentration (LC_{50}):

The assessment of the LC_{50} has been used as an important parameter to measure toxicity and screen the toxicity effect of a substance. In the current study, the LC_{50} of Ag-NPs for African catfish (*C. gariepinus*) was determined to be 171.7 mg/l. Meanwhile, **Hamed Abdel-Tawwab (2021)** found that the LC_{50} of Ag-NPs in Nile tilapia was 20.0 mg/l. The above observation was supported by **Chakeri *et al.* (2015)** they noted that heavy metal poisoning causes anxiety and eventually death and they recorded that different fish species, had different sensitivity to heavy metals. Variability in toxicity depends on age, size and condition of the tests **(Rathore and Khangarot., 2002).**

In this investigation LC_{50} values indicated that Ag-NPs were toxic to studied fish. This result was explained by **Wu *et al.* (2010)** who suggested that Ag-NPs have cytotoxic and harmful effects in fish which also, agree with **(Yalsuyi and Vajargah 2017 and Kumar *et al.*, 2018).** Previous studies recorded high toxicity of Ag-NPs to some species such as *Corophium volutator* (**Chen and Folt, 2000 and Belitz *et al.*, 2001).** The variations in LC_{50} values may be due to the difference in fish species. Also, **Barbieri and Fereira, (2011)** explained the difference in LC_{50} values could be attributed to several factors including inter- laboratory variations or different strains. Also, differences in the LC_{50} values among the same fish species depend upon the age, body weight, sex, water temperature and feeding condition of the experiment **(Hedayati *et al.*, 2012).**

Clinical and behavioral change:

Behavioral responses are important tool in aquatic toxicology where fish noted behavioral change as signs of toxicity **(Ghazala *et al.* 2014).**

Clinical signs and the postmortem findings of African catfish *(C. gariepinus)* are associated with toxicity of 50 mg/L Ag-NPs after 30 days were revealed by abnormal skin pigmentation in the form fading and abnormal swimming behavior in the form of erratic swimming. loss of equilibrium and abnormal swimming were caused by deficiencies in nervous and muscular coordination, which may be due to accumulation of acetylcholine at synaptic junctions **(Prasanth *et al.*, 2005)**. Abnormal skin pigmentation may be due to exposed to Ag-NPs. Where the pigmentation in fish is nervously controlled. These results supported by **Sayed (2016)** who recorded change behavior in

African catfish (*C. gariepinus)* exposed to toxic 4- nonylphenol. Some research suggests that the toxicity of Ag-NPs depend upon many factors such as size, shape, surface charges and surface area **(Hedayati *et al.*, 2012)** or toxicity has been suspected to arise in part from the release of Ag+ ions.

This study supported with **Mekkawy *et al.*, (2011); and Campos-Garcia *et al.*, (2016)**. From this study we noticed that co-supplementation of catfish with NS declined the abnormal behavior which caused by the exposed of fish to Ag-NPs. This may be due to its protective effects **(Mohammed and Arias, 2016)**

Biochemical parameters:

Serum ALP, ALT and AST level:

Biochemical studies have been used for assessing the health status, natural status and adaptation to the environment (**Abdel-Tawwab and Wafeek, 2017**). Assessment of ALP, ALT and AST levels in serum are perfect indicators for diagnosis of liver damage in animals exposed toxic nanoparticles **(Samipillai and Jagadeesan, 2005)**. This support the present result which stated that the Levels of ALP, ALT, and AST of catfish exposed to Ag-NPs increased significantly after 30 days of exposure to 50 mg/L Ag-NPs.

Also, **Imani *et al.* (2015)** recorded a significant increase in the activities of ALP, ALT and AST of rainbow trout (*O. mykiss*) after exposed to Ag-NPs. In agreed with this results **Abdel-Khalek *et al.* (2015)** found increase in serum activities of ALP, ALT and AST of Nile tilapia *(O. niloticus*) after exposed to copper oxide nanoparticles. Increasement in ALT, ALT and AST levels could be due to hepatic cell injury or increased synthesis of the enzymes by the liver. These increases supported by **Hasan *et al.* (2018) and Kovacik *et al.* (2019**) they attributed this increase to different of conditions, including intestinal and hepato-pancreatic injury, muscle damage and toxic hepatitis**.** Some authors associated between increase of hepatic enzymes and liver tissue damage **(Eminzade *et al.*, 2008). Kavitha *et al.* (2010)** recorded an increase in liver enzymes in an Indian major carp (*C. catla*) after exposure to arsenate. High levels of ALP, ALT and AST in the serum could a result of liver injury leading to increased permeability of cell membrane.

Atli *et al.* (2015) explained the increase of these enzymes into the bloodstream due to a damage occurred in the hepatic tissues of catfish exposed to Ag-NPs due to the accumulation of Ag-NPs in liver tissues. This results are in accordance with **Naguib *et al.*, (2020) and Hamed *et al.*, (2021),** They attributed this to induced destroy and degeneration of hepatocytes as a result of Ag-NPs exposure. Other results recorded decrease in serum ALP, ALT and AST **(Hedayati *et al.*, 2019)** could be attributed to multiple factors: different concentration of Ag-NPs and different fish species.

Serum urea, creatinine and uric acid:

Serum (Urea, Creatinine and Uric acid) used as indicator to glomerular filtration rate and kidney dysfunction **(Maita *et al.*, 1984),** moreover indicate to several disturbances in the fish excretory kidney (**Maxine and Benjamine, 1985**).

As showed in this study the results of urea, creatinine and uric acid were significantly increased after exposed to 50 mg/L Ag-NPs p for 30 days. These results agree with **Sayed and Hamed (2017) and Abdel-khalek *et al.* (2015)** They found that exposure of 4-nonylphenol caused increase in serum urea, creatinine and uric acid in African catfish (*C. gariepinus)*. These results supported by **Abdel-khalek *et al.* (2015)** recorded that increase serum urea and creatinine in African catfish (*C. gariepinus*) after exposure to carbofuran. They attributed this elevation in renal function to the renal dysfunction and reduced glomerular filtration rates or might be due to renal dysfunction as shown by decrease glomerular filtration rates **Hamed and Osman, 2017 and Abu mourad *et al.*, 2005** they recorded increase kidney function in African catfish (*C. gariepinus*) after exposed to carbofuran and glyphosate, respectively. The dietary NS diet significantly enhanced the renal products parameters of Ag-NPs exposed fish. The results demonstrated that NS has hepato-renal protective effects against Ag-NPs toxicity

Present results disagreed with **Mekkawy *et al.* (1996)** who noted that serum urea was not change in Nile tilapia *(O. niloticus*), which were supported with **Haggag (2004)** who recorded decrease kidney function in Nile tilapia (*O. niloticus*) after exposed to bisphenol A. This may be attributed to different fish species or exposure chemicals.

Serum protein, albumin and globulin level:

Proteins play vital role in the regulation of water balance and metabolism of living beings (**Bakhshwan *et al.*, 2009)** and play an important role in the maintenance of osmotic balance between the circulating blood and the tissue membrane **(Kadry *et al.*, 2012)**.

Results of total protein, albumin and globulin showed a significant ($P \leq 0.05$) increase in exposed of 50 mg/L Ag-NPs However, A/G ratio decreased. In the same respect **Manjula *et al.* (1996)** who recorded that increase in total protein level in red swamp crayfish (*P. clarkia*) after exposure to cadmium. Also, the increase of total protein supported by **Öner *et al.* (2008)** and **Vajargah *et al.* (2019)** in Nile tilapia (*O. niloticus* after exposed to Ag. They indicated that the increase of total protein is attributed with the liver damage which lead to a liver dysfunction and damage of hepatocytes. The results are in consistent with **Marie *et al.* (1998)** who recorded marked increase in serum total protein level of common carp exposed to profenofos. This is agreement with **Feng *et al.* (2000)** reported that Ag-NPs may penetrate the cell wall affecting protein synthesis.

Serum glucose level:

Serum glucose is considered as indicator for stress response in fish **(Abdel-Baky, 2001).** As shown in the present study the serum glucose recorded a significant increase in exposed of 50 mg/L Ag-NPs on African catfish. These results are agreement with **Farmen *et al.* (2012)** who recorded increase in serum glucose level of juvenile Atlantic salmon (*samo salar*) after exposed to Ag-NPs. This is in similar to **Shaluei *et al.* (2013)** who found increase in serum glucose level of silver carp *(Hypophthalmicthys molitrix*) after exposure to Ag-NPs. This author stated that this increase caused due to the hyperglycaemic condition cause by environmental pollution might produce stress on fish by the inhibition of the gluconeogenesis process which elevate glycogen breakdown and increase serum blood glucose level.

In the same line **Abdel-khalek *et al.* (2015)** found a siginficant increase in serum glucose level of Nile Tilapia (*O. niloticus)* after exposure to copper oxide nanoparticles. This is supported by

Wang *et al.* (2008) and Hedayati *et al.* (2019) They recorded that the alteration in serum glucose level may be related to the liver or kidney damage, glycogenolysis, nutritional deficiencies and glucose synthesis The hyperglycaemic can also caused by inhibition of cholinesterase at neoro-effector site in the adrenal medulla, leading to hypersecretion of adrenaline and cortisol which stimulates the breakdown of glycogen to glucose.

In addition, **Reddy and Leatherland (1998)** showed increased breakdown of muscle and liver glycogen due to reduced insulin activity as result of the hurt of the pancreatic cells which lead to increase glucose level. These elevations in glucose levels could be attributed to the raise in glucogenolysis process in stressed fish to satisfy its energy needs **(Hamed and Abdel- Tawwab, 2017; Abdel-Tawwab and Hamed, 2018)**. But, in different with **Kovacik *et al.* (2019)** who recorded decrease in glucose levels in male goldfish (*C. auratus gibelio*) after exposed to Ag-NPs.

Serum cortisol level:

Cortisol acts as indicator for stress response in African catfish *(C. gariepinus*) and increase with pollution and heavy metals According to **Abdel-Baky (2001).**

Serum cortisol is the major corticosteroid hormone in fish and may have a significant effect on its dynamics **(Mommsen *et al.*,**

1999). Concerning serum cortisol, the present results showed a significant increase in the exposed catfish of 50 mg/L Ag-NPs after 30 days. This increase in cortisol which noted from the present result was agreed with **Monfared and Soltani (2013)** who recorded increase serum cortisol level of rainbow trout *(O. mykiss*) after exposure to Ag-NPs. This is supported by **Kadry *et al.* (2012**) who found increase in serum cortisol level of African catfish (*C. gariepinus)* after exposure to carbofuran. The increased in cortisol levels where attributed to the activation of hypothalamo-pituitary inter renal axis with release cortisol in blood stream due to stress.

In the same manner **Reddy and Leatherlend. (1998)** who demonstrated that the activation of hypothalamus – pituitary – inter renal axis with their release of steroid cortisol in blood stream due to stress and the elevation in osmotic water – influx, induced increase cortisol levels which cause different physiological changes. This disagreed with **Maduenho and Martinez (2008)** who observed no change in plasma cortisol level exposed to diflubenzuron in (*P. Lineatus*). This different because varied fish species or exposure period.

Serum LDH level:

Lactate dehydrogenase (LDH) enzyme is predominated in the liver **(Kori-Siakpere *et al.*, 2012)** used as indicator of tissue damage **(El-Gohary and Marwa 2011)**. When damage occurs to liver tissues these enzyme is released into the blood stream lead to increased level in plasma **(Atli *et al.,* 2015).** This is agreement with the present results which indicated increase in serum LDH enzyme in exposed of 50 mg/L Ag-NPs. this results are supported by **Das *et al.* (2004)** who stated that this increase may be due to the incidence of anoxia throughout tissue damage or stress conditions. This is agreed with **Hedayati *et al.* (2019)** who recorded increase LDH level in Zebrafish (*D. rerio*) after exposed to silver nanoparticle (Ag-NPs) and Silver nitrate ($AgNO_3$). And agree with **Abdel El-Atti *et al.* (2020)** who noted that exposed to Ag-NPs caused increase in serum LDH level of crayfish (*P. clarkii*).

The elevation in LDH may be due to plasma membrane damage, changes in the energy production and cellular metabolism pathway during Ag-NPs exposure (**Domingues *et al.*, 2010**). These results disagree with **Abhijith *et al.* (2016)** who recorded that LDH level decreased in liver of Catla carp (*C. catla*) after exposed to methyl parathion. Also, **Sayed and Soliman (2017)** found decrease in serum LDH level in African catfish (*C. gariepinus)* after exposure to silver nanoparticles.

Serum cholesterol and triglycerides level:

Additionally, present study recorded an increase in serum lipid levels (Cholesterol and Triglyceride) demonstrated in African catfish *(C. gariepinus*) which exposed to 50 mg/L Ag-NPs this results are in agreement with **Valerio-Garcíaa *et al.* (2017) and Yang and Chen (2003)**. They recorded increase in total lipid levels of goodie fish (*Chapalichthys pardalis)* exposed to silver nanoparticles. Increase in serum total lipids may be due to disturbance of lipid metabolism **(Shalaby, 2001).** In supporting to the previous results **Öner *et al.* (2008)** who recorded that the liver and kidney stress and failure circumstance causing the release of cholesterol into the blood stream

Vajargah *et al.* (2019) demonstrated that Cholesterol serum level is used as an indicator of liver dysfunction; it's the most important sterol occurring in plasma which is known to take part in elevation total lipid. Also, who recorded an increase in serum cholesterol level of goldfish (*C. auratus gibelio*) after exposed to Ag-NPs. Other result by **Kovacik *et al.* (2019)** showed a significant increase in cholesterol level of *Cyprinus carpio* after exposed to trace metals. The increment in cholesterol level could be attributed to excess energy reserves which are required by stressed fish as a response to carbofuran intoxication **(Hamed and Osman, 2017)**.

Similarly, **Öner *et al.*, (2008)** showed increase levels in serum triglyceride after exposed to metal (Ag, Cd, Cr, Cu and Zn) in Nile tilapia (*O. niloticus*) higher triglyceride level may occur with nephritic syndrome and glycogen storage impairment. these results are disagreed with **Vajargah *et al.* (2019)** who recorded no change in triglyceride level after exposed to silver nanoparticles in goldfish (*C. auratus gibelio)*. This variant may be attributed to the different fish species or exposure period.

The present results are in agreement with **Farmen *et al.* (2012)** who recorded that Ag-NPs at 100 mg^{-1} concentrations in water increased gill accumulation of Ag in Atlantic salmon and result in numerous biochemical and physiological responses indication of toxicity.

Oxidative stress and antioxidant enzyme:

Antioxidant enzymes can indicate the antioxidant status of the organisms, moreover they are useful biomarkers reflecting to exposure to pollutants and their toxicity **(Valavanidis *et al.* 2006).** Several environmental toxicities related to oxidative stress and caused oxidative damage **(Velisek *et al.*, 2011)**.

Liver and kidney LPO content:

The present results recorded an increase in LPO level in exposed catfish to 50 mg/L Ag-NPs after 30 days. It was also showed that Ag-NPs toxicity cause a remarkable change in hepatic and renal oxidative stress. In agreement with the present results **(Hunag *et al.,* 2003)** reported that lipid peroxidation (LPO) has been showed the main contributor to the loss of cell function under oxidative stress. **Strużyński *et al.* (2013)** recorded an increase in LPO activity in spinycheek crayfish *(O. limosus)* after exposure to Ag-NPs and increase in free radicals. This could be explained by the elevation in free radicals and overproduction of ROS due to oxidative stress. Likewise, the responses of LPO may different with time and concentration of toxicants **(Benavides *et al.*, 2016).** The elevated enzymatic activities might have reflected the better protection against toxicity induced lipid peroxidation **(Abdel-Tawwab and Wafeek 2017).**

In other investigations, **Sayed and Hamed, (2017)** reported that levels of LPO significantly decreased in African catfish *(C. gariepinus*) exposed to 4-nonylphenol and decreased in tilapia fish (*O. niloticus*) exposed to pendimethlin-intoxicated, respectively.

Liver and kidney GSH content:

Reduced glutathione (GSH) plays a vital role in preventing the oxidative damage and maintaining cellular redox status **(Slatinskã *et al.*, 2008).**

The present results demonstrated an increase in GSH content in exposed catfish to 50 mg/L Ag-NPs after 30 days. Overutilization of existing enzymes to overcome oxidative stress caused by silver nanoparticles, which caused elevation in GSH content in specific organs **(Federici *et al.*, 2007).**

The increase in GSH content which recorded from this study is in agreement with **(Hamed and Osman 2017)** who found increase in GSH in African catfish (*C. gariepinus)* after exposed to carbofuran. This is supported by **Zhang *et al. (*2004)** and **Abd El-Atti *et al.* (2020)** they recorded increase in GSH content of crayfish (*P. clarkii*) after exposed to Ag-NPs. Elevation in GSH level indicated its adaptive and protective role against oxidative stress. According to **(Amira and Shiba 1992)** GSH enzyme protect the membrane-polyunsaturated fatty acid from free radical and lipid peroxidation by abstracting the hydrogen of sulfohydryl group instead of methyl hydrogen of unsaturated lipids.

whereas **Rajkumar *et al.* (2015)** recorded that the decrease in GSH content of Rohu (*L. rohita*) after exposed to silver nanoparticles may attributed to varied tested chemicals and different exposure periods.

Liver and kidney CAT content:

CAT content provide the first line of defense against reactive oxygen species and used as biomarker of oxidative stress **(Abdel-**

Tawwab and wafeek 2017). The main function of CAT is to catalyze the decomposition of H_2O_2 **(Chelikani *et al.,* 2004).** The results in this study demonstrated an increase in liver and kidney CAT content in catfish exposed to Ag-NPs. In agreement with these results **Jayaseelan *et al.* (2014)** who noted increase in CAT content of Mozambique tilapia (*O. mossambicus)* after exposed to Nickel nanoparticles. In the same trend **Sayed *et al.* (2016)** recorded increase in CAT content of African catfish *(C. gariepinus)* after exposed to 4-nonylphenol. In support of the present result **Abd El-Atti *et al.* (2020)** reported an increase in the CAT content of crayfish (*P. clarkii*) after exposed to silver nanoparticles.

The increase of CAT activities may be an adaptive mechanism to prevent the accumulation of toxic reactive oxygen intermediates **(Regoli *et al.,* 2006).** Several author observed that the increase in the CAT enzyme caused by converts oxygen free radical (O_2^-) into $H2O_2$ and then into H2O and O2 **(Jin *et al.,* 2010).** In addition, **(Moraes *et al.*, 2009)** explained increase in CAT enzymes caused by the flux of superoxide radicals due to the oxidative stress caused by toxic exposure.

Kidney and liver SOD content:

SOD is an enzyme that primarily catalyzes the dismutation of superoxide radicals, induced in tissues that are suffering from oxidative stress **(Sakai *et al.*, 1992).** SOD plays an important role in the defense against the toxic effect and maintain the balance between oxidant and antioxidant **(Li *et al.*, 2003).**

From the present study that exposed catfish to 50 mg/L Ag-NPs enlarged in SOD content, similar result has also been noted by other authors **Benavides *et al.* (2016)** who studied increase in SOD activity in *Carassius auratusafter* after exposure to aluminum (Al_2O_3) and (ZnO) oxide nanoparticles, this increase in SOD activity as the defense mechanism against oxidative stress in several aquatic organisms. This is agreement with **Hamed and Osman (2017)** who found increase in SOD activity in liver and kidney African catfish *(C. gariepinus*) after exposed to carbofuran. In supporting to these results **Abou-Zeid and Khalil, (2014)** recorded Increase in SOD activity may be attributed to high production of superoxide anion radical.

Disagreement of these results showed by **Faheem *et al.* (2020)** who recorded decreased in *Labeo rohita* larvae after exposure to BPA.

Liver and kidney TAC:

It was showed that African catfish *(C. gariepinus*) exposed to 50 mg/L Ag-NPs had increased total antioxidant capacity (TAC). This is because the liver is major sensitive organ of the body. Presence of transition metals and the metallic nature of nanoparticles induce the production of ROS, causing oxidative stress (**Rajkumar *et al.,* 2015**). The present study was supported by **Abdel-wahhab *et al.* (2016)** who recorded an increase in TAC of African catfish *(C. gariepinus)* after exposure to chitosan nanoparticles.

In the same line **Abou Khalil *et al.* (2017)** found increase in TAC in African catfish (*C. gariepinus*) after exposed to 4-nonylphenol, this increase induced because mounting evidences indicated shift of redox state towards pro-oxidant side under stress by induction of depletion of enzymatic and nonenzymatic antioxidant and reactive oxygen species formation **(Kourouma *et al.,* 2015)**. This disagree with **Sayed *et al.* (2016)** who studied that exposed to 4-nonylphenol caused decrease in TAC of African catfish (*C. gariepinus)*.

DNA fragmentation:

To evaluate the effects of Ag-NPs on African catfish (*C. gariepinus*) DNA fragmentation was analyzed. The obtained data demonstrated that Ag-NPs increased DNA fragmentation and these results agree with that obtained by **(Kim *et al.,* 2014).** Oxidative stress and overproduction of free radicals caused genotoxicity (**Abdel-Wahhab *et al.,* 2016 and Sayed and Hamed., 2017**). Moreover, silver nanoparticles elevated the DNA damage in the length of tail **(Cong *et al.*, 2011). Hossain and Hug. (2002)** demonstrated that Ag^+ ions bind with DNA by covalent bonds and caused DNA damage this mechanism explains the elevation of DNA fragmentation in fish.

More investigator showed the effect of Ag-NPs in hepatic and renal DNA **Gagné *et al.*, (2013)** showed effect of silver nanoparticles on tilapia fish. Other study supported by **Taju *et al.* (2014)** found change in DNA in *C. catla* and *labeo rohita* after exposed to silver nanoparticles. **Sayed (2016)** recorded elevation in African catfish (*C. gariepienus)* in DNA fragmentation after exposed to silver nanoparticles. Degeneration after exposed to Ag-NPs increased, this may be as genotoxicity or DNA damage. This result is agreement with **Feng *et al.* (2000)** who found that Ag-NPs may be penetrate the cell wall affecting DNA change. Also, **Gagné *et al.* (2012)** has been showed that exposure of trout to Ag-NPs led to a 62% drop in DNA break. It has been found that exposure of human blood cell with Ag-NPs resulted in DNA break and they stated the DNA damage was resulted from necrosis or apoptosis and cytotoxicity **(Flower *et al.*, 2012).**

This study recorded increase DNA damage in African catfish exposed to 50 mg/L Ag-NPs. Whereas, DNA fragmentation was declined in group which exposed to 50 mg/L Ag-NPs and 3% NS seeds. Stimulatingly, Dietary NS reduced DNA damage in hepato-renal tissues. This could be explained by binding of NS to DNA at sites that would normally react with the active metabolites of carcinogen. Therefore, the results suggest that NS possess DNA-protective properties **(Abd-Elkareem *et al.*, 2020)**

Histological changes:

Liver histopathological changes:

Histopathology alterations in liver catfish are important biomarkers in toxicological studies **(Singh *et al.*, 2018)** recorded that histopathology of the liver can be used to show effect of aquatic animals which exposed to toxic materials.

Histological studies by several investigators were found a useful tool for assessing the damage of tissues caused by different nanomaterials **(pournori *et al.*, 2017).** Different degree of damage especially in the liver showed enlargement and kidney hemorrhage in the exposed fish *C. auratus* to aluminum (AL_2O_3) and zinc (ZnO) oxide nanoparticles. The observed signs might be attributed to the stress condition associated with the toxicant exposure **(Benavides *et al.*, 2016)**. According to **Johari *et al.* (2015)** who stated that tissue damage in *O. mykiss* might be due to exposed to Ag-NPs. Also, auther found increase pigmentation in the liver tissues which was likely related to pigmented macrophages.

From the present study it was recorded that histological analysis of liver tissue of African catfish *(C. gariepinus)* exposed to 50 mg/L Ag-NPs after 30 days was showed histopathological response to organic compounds. This is agreement with **Jayaseelan e*t al.* (2014)** who observed effect of nickel nanoparticles on Zambique tilapia (*O. mossambicus*), the main alterations are irregular shaped nuclei, cytoplasmic vaculation, pyknotic nucleus, nuclear degeneration, nuclear hypertrophy and necrosis. This is supported by **Brusle and Anadol (1996)** who explained that the Liver is an important organ in active metabolism, detoxification and extremely sensitive to pollutants.

Other results recorded effect of silver nanoparticles on zebrafish (*D. rerio)* (**Yazdanparast *et al.*, 2016**). In supporting with **Freeman *et al.* (1983)** who demonstrated that liver of Atlantic cod (*Gadus morhae*) particularly is the principle organ of detoxification so, the increases in ALP, ALT and AST in the liver enzymes indication of damage in the exposed fish these due to the induced of necrosis or hepatitis of cells.

In this study the histopathological changes in liver noted hepatocytes with marked hydropic degeneration, multiple foci of spotty necrosis, portal tracts showing minimal inflammatory cellular, mild interface hepatitis and fibrosis. similar **Ostaszewska *et al.* (2018); and Naguib *et al.* (2020)** they observed proliferation of hepatocytes, cytoplasmic vaculation, infiltrations of inflammatory cells, hepatic necrosis, and dilation in the blood vessel of *Labio rohita*, rainbow trout (*O. mykiss)* and African catfish (*C. gariepinus)* (Naguib *etal.,* 2020) exposed to Ag-NPs, respectively. These alterations may be attributed to the toxic effect of nanoparticles on liver cells.

The toxic effects of Ag-NPs were observed in rainbow trout hepatocytes **(Frakas *et al.,* 2011)** who founded Changes of genetic materials such as nuclear hypertrophy, karyolysis, necrosis and Reduction in membrane integrity primary hepatocytes. More studies using Ag-NPs showed change in gill, skin and liver in Mozambique tilapia (*O. mossambicus*) **(Govidasamy and Abdul-Rahuman, 2012).**

Kidney histological changes:

Pollutants to fish caused inflammatory cells, displacement of nuclei, shrinkage of glomeruli, disintegration of the renal epithelium and breakdown of bowman's capsules caused change in kidney function **(Al-Zahaby *et al.*, 1998).**

The histopathological alteration that were noted among the kidney tissue in catfish exposed to Ag-NPs showed obsolescent glomeruli with mild inflammatory cellular infiltrate, vascular and hydropic degeneration in tubules, necrosis. It was also show a reduction in the size of the glomular vessels, change in glomerular cell, and Bowman's space due to accumulated edema, increase fluid movements from glomular pool. These histopathological changes may be due to toxic and accumulation of nanoparticles on cells and tissues.

These results were in agreement with **Wu and Zhou (2013)** and **Kotb *et al.* (2018)** they showed nephrotoxicity in *C. gariepinus* was characterized by disturbances in blood parameters specific for kidney function like creatinine and uric acid after exposed to 4- nonylphenol. In the same line **(Abu mourad *et al.*, 2005)** found kidney interstitial hemorrhage, tubular degeneration and glomerular edema.

Dietary NS plays modulatory role in preventing and /or repairing the histopathological abnormalities of liver and kidney tissues induced by Ag-NPs **(Kotb *et al.*, 2018)**

Nigella sativa:

Nigella sativa seeds have been widely used as a remedy for several diseases in medicine **(Ijaz *et al.*, 2017).**

From the present study it was observed that increase in liver enzymes and kidney function in catfish expose to Ag-NPs, indicating to the hepato-renal injury. These results may be attributed to Ag-NPs accumulation and deposition in fish tissues. As expected, feeding fish with NS diet reduced liver enzymes (ALT, ALP and AST) and kidney function (urea, creatinine and uric acid) levels, demonstrating it protective effect against Ag-NPs. In similar study **Farooqui *et al.,* (2017)** who noted the effect of *Nigella sativa* seeds on kidney of rat.

NS has numerous beneficial properties, including its ability to improve antioxidant parameters (LPO, SOD and CAT) Against fipronil in rats **(Abdel-Daim *et al.*, 2018).**

NS supplementation has been shown to reduce disease such as fever, common cold, rheumatic diseases and less inflammatory tissue **(Al-Jishi 2000 and Muhammad and Al-Ghamdi 2002). Kotb *et al.* (2018)** recoded that NS can improve liver and kidney function in *C. gariepinus* after exposed to 4-nonylphenol. Moreover, decrease biomarker parameters such as total protein, albumin, globulin and cholesterol levels after receiving NS diet **(Diab *et al.*, 2004; Al-Kubaisy and Al-Noaemi, 2007 and Aly *et al.*, 2019).**

NS seeds used as a natural treatment for many aliments, most investigation confirms its good effect on biochemical results and value in anti-oxidant, anti-inflammatory **(Salah *et al.*, 2019).** In the same line **Kanter *et al.* (2005)** recorded that *Nigella sativa* have positive effect to clear enhancement in the antioxidant in rat after exposed to cadmium. Other investigator showed an improvement of NS on kidney of African catfish (*C. gariepinus*) after exposed to Aeromonas hydroppila Vaccine **(Aly *et al.*, 2019).**

Conclusion:

In conclusion, based on the results obtained from the present study, it was found that exposure to 50 mg/L Ag-NPs and/ or NS produced significant modulator effects on the varies behavioral and clinical changes of catfish which presenting as morphological alterations after exposed to Ag-NPs.

The decline in the liver function (ALP, ALT and AST) and kidney function (Urea, creatinine and uric acid) under the effect of Ag-NPs and /or NS might have been the result of a direct action on a damage occurred in the hepatic and renal tissues. In addition, the decrease in total protein, glucose, cortisol level and LDH by Ag-NPs and NS administration may be related to the raise in glucogenolysis and liver damage.

The current study showed that administration of NS stimulated the total lipid, oxidant and antioxidant of catfish and improved them to levels similar to the control group. It is also concluded that NS reduced DNA damage and histopathological changes in hepatic and renal tissues.

Further studies are still required to understand the toxicological mechanisms of Ag-NPs in fishes so that coming studies should shed new lights on the important activity of Ag-NPs ultimately biochemical, histo- pathological and other adverse impacts.

SUMMARY

Effect of *Nigella sativa* on silver nanoparticles induced hepato-nephrotoxicity in *Clarias gariepinus*

The present work has been planned to investigate the effect of Ag-NPs on selected biochemical parameters of serum, liver tissue and kidney tissue in *Clarias gariepinus*. This study was carried out also to evaluate the effect of Ag-NPs on the oxidative stress, antioxidant enzyme and DNA fragmentation. In addition, this study aimed to highlight the histopathological change to Ag-NPs exposure. Finally, to represent the role of NS in the improvement of these parameters.

Experimental design and investigated parameters:

In the present study 168 catfish (*C. gariepinus*) were used. 48 fish were used to estimate the (LC_{50}) of Ag-NPs. The other 120 fish were divided into four groups and each group consisted of 3 replicates of 10 fish per aquarium. Group I was served as control group. Group II served as control group and co- administrated with 3% NS. Group III was exposed to 50 mg/L Ag-NPs. Group IV was exposed to 50 mg/L Ag-NPs and co- administrated with 3% NS.

After 30 day of treatment, blood samples were collected from the caudal blood vessels. Serum was collected, liver and kidney were rapidly excised for the biochemical and histological studies.

The present work included the following investigations:

- Clinical investigation and post mortem examination

- Serum: Determination of serum levels of ALP, ALT, AST, urea creatinine, uric acid, total proteins, Albumin, Globulin, glucose, cortisol, LDH, cholesterol and triglycerides.

- Liver and kidney determination of LPO, GSH, CAT, SOD and TAC content.

- DNA fragmentation analysis.

- Liver and kidney histopathological changes.

The results:

The half lethal concentration (LC_{50}) of Ag-NPs was observed through exposure of catfish *(C. gariepinus)* was 171.7 mg/L in 96 h.

Clinical behavior and biochemical parameters:

The clinical signs of *C. gariepinus* associated with toxicity of 50 mg/L Ag-NPs were manifested by nervous signs including abnormal swimming behavior in the form of irregular swimming with or without rove, abnormal skin pigmentation in the form of wilt increase mucous secretion and rapid dill movement. The postmortem findings exposed, different degrees crowding especially in all internal organs of the exposed fish. As well as the enlargement of liver. The biochemical analysis revealed an increase in liver enzymes values (ALP, ALT and AST) in addition to urea, creatinine, uric acid, total protein, albumin, globulin glucose, cortisol, LDH and total lipid levels were also detected in group (III) exposed to 50 mg/L Ag-NPs for 30 days.

Oxidative stress parameters in liver and kidney tissues recorded that:

The exposure of 50mg/L Ag-NPs in African Catfish *(C. gariepinus)* led to increase in liver and kidney lipid peroxidation

(LPO), reduced glutathione (GSH), catalase enzyme (CAT), superoxide dismutase (SOD), total antioxidant capacity (TAC) these compared to control group and treatment group II 3% *NS*.

DNA in liver and kidney tissues:

The result of exposed 50 mg/L Ag-NPs in liver and kidney DNA damage increase in exposed Ag-NPs group III compare to control group and treatment group II with 3% NS.

The most observed histopathological alterations included:

-Liver: Hepatocytes with hydropic degeneration, portal tract showing inflammatory cellular infiltrate, mild fibrosis, interface hepatitis and shows multiple foci of spotty necrosis.

-Kidney: alternation in glomeruli with inflammatory, degenerative change of renal tubules and showed inflammatory cellular infiltrate.

Conclusion:

Feeding of *C. gariepinus* on a NS enriched diet minimized the side effects of Ag-NPs toxicity and improve the above- mentioned variable near to the control group. The current study has revealed that dietary NS has good effect against Ag-NPs toxicity in African catfish (*C. gariepinus*).

REFERENCES

Abd El-Atti, M.; Desouky, M.; Mohamadien, A. and Said, R. (2020). Sub-lethal toxicity of silver nanoparticles on crayfish, *procambarus clarkii*: oxidative stress, DNA damage, and histopathological alterations. Arct. J., 73(3): 79-108.

Abdel-Baky, T. E. (2001). Heavy metals concentrations in the catfish, *Clarias gariepinus* (Burchell, 1822) from River Nile, El-Salam canal and Lake Manzala and their impacts on cortisol and thyroid hormones. Egypt. J. Aquat. Biol. Fish., 5(1) :79-98.

Abdel-Daim, M. M.; Shaheen, H. M.; Abushouk, A. L.; Toraih, E. A.; Fawzy, M. S.; Alansari, W. S.; Aleya, L. and Bungau, S. (2018). Thymoquinone and diallyl sulfide protect against fipronil- induced oxidative injury in rats. Environ. Sci. Pollut. Res. Int., 25 (24): 23909- 23916.

Abd-Elkareem, M.; Khalil, N. S. A. and Sayed, A. E. D. (2020). Cytoprotective effect of *Nigella sativa* seed on 4-nonylphenol-induced renal damage in the African catfish (*Clarias gariepinus*). Chemosphere, 259:127-379.

Abdel-Khalek, A. A.; Kadry, M. A. M.; Badran, S. R. and Marie, M. A. S. (2015). Comparative toxicity of copper oxide bulk and nano particles in Nile Tilapia, *Oreochromis niloticus*: biochemical and oxidative stress. J. Basic. Appl. Zool., 72: 43-57.

Abdelrazek, H. M. A.; Kilany, O. E.; Muhammed, M. A. A.; Tag, H. M. and Abdelazim, A. M. (2018). Black seed thymoquinone improved insulin secretion, hepatic glycogen storage, and oxidative stress in streptozotocin- induced diabetic male wister rats. Oxid. Med. Cell. Longev., 81:41-65.

Abdel-Tawwab, M. and Hamed, H.S. (2018). Effects of bisphenol A toxicity on growth performance biochemical variables, and oxidative stress biomarkers of Nile tilapia, *(Oreochromis niloticus*) (L.). J. Appl. Ichthyol., 34(5): 1117-1125.

Abdel-Tawwab, M. and Wafeek, M. (2017). Fluctuations in water temperature affected waterborne cadmium toxicity: hematology, anaerobic glucose pathway, and oxidative stress status of Nile tilapia, *Oreochromis niloticus* (L.). Aquaculture, 477: 106-111.

Abdel-Wahhab, M. A.; Aljawish, A.; Kenawy, A. M.; El-Nekeety, A. A.; Hamed, H. S. and Abdel-Aziem, S. H. (2016). Grafting of gallic acid onto chitosan nanoparticles enhances antioxidant activities *in vitro* and protects against Ochratoxin A toxixity in catfish (*Clarias gariepinus*). Environ. Toxicol. Pharmacol., 41: 279-288.

Abhijith, B. D.; Ramesh, M. and Poopal, R. K. (2016). Responses of metabolic and antioxidant enzymatic activities in gill, liver and plasma of *catla catla* during methyl parathion exposure. J. Basic. Appl. Zool., 77:31-40.

Abou Khalil, N. S.; Abd-Elkareem, M. and Sayed, A. H. (2017). *Nigela sativa* seed protects against 4-nonylphenol-induced haematotoxicity in *Clarias gariepinus* (Burchell, 1822): oxidant/ antioxidant rebalances. Aquacult. Nutr.,1-8.

Abou-Zeid, E. H. and Khalil, S. A. (2014). Effect of acute fenitrothion insecticide exposure on DNA damage and oxidative stress biomarkers and health of Nile tilapia fingerlings, *(Oreochromis niloticus)* L. World J. Fish. Mar. Sci., 6: 361-370.

Abu mourad, I. M. K.; El-Maedawy, S. A. and Kenawy, A. M. (2005). Toxicity evaluation of glyphosate herbicide on catfish, *(Clarias gariepinus*). Egyptian. J. Exp. Biol., 1:87-94.

Aebi, H. (1984). Catalase *in vitro*. Methods Enzymol., 105:121-126.

Al-Jishi, S. A. A. (2000). A study of *Nigella sativa* on blood hemostatic functions. M. Sc. Thesis [King faisal University, Dammam, Saudi Arabia].

Alkaladi, A.; Nasr, A. M.; El-Deen, N.; Afifi, M. and Abu Zinadah, O. A. (2015). Hematological and biochemical investigations on the effect of vitamin E and C on *Oreochromis niloticus* exposed to zinc oxide nanoparticles. Saudi. J. Biol. Sci., 22: 556-563.

Al-Kubaisy, K. and Al-Noaemi, M. (2007). A protective role of *Nigella sativa* oil against the harmful effect of CCl4 on the liver cells. Int. J. Nutr. Wellness, 3:1-14.

Aly, S. M.; Ismail, M. M.; Fathi, M. and Al Zohairy, M. A. (2019). The role of *Nigella sativa* in improving the immune response of the African catfish (*Clarias gariepinus*) to Aeromonas hydroppila Vaccine. Egypt. J. Aquat. Biol. Fish., 23 (4) :373-384.

Al-Zahaby, A. S.; Hemmaid, K. Z.; Gamal, A. M. and Ghonemy, O. I. (1998). The pollutant effects of copper, zinc and lead on the histological patterns of fish kidney. Egyptian J. Aquat. Biol. Fish., 2 (3) :15-41.

Amin, B. and Hosseinzadeh, H. (2016). Black cumin (*Nigella sativa)* and its active constituent, thymoquinone: an overview on the analgesic and anti-inflammatory effects. Planta. Med., 82 :8-16.

Amira, R. and Shiba, R. (1992). Radioprotective effect of exogenous glutathione on rat parotid glands. Int. J. Radiat. Biol., 61 (5) :595-702.

Amlacher, E. (1970). "Text book of fish diseases " pp:135-137. T. E. S. publication, 1970, Jercy, USA.

Arora, S., Jain, J., Rajwade, J. and Paknikar, K.; (2009). Interactions of silver nanoparticles with primary mouse fibroblasts and liver cells. Toxicol. Appl. Pharmacol., 236 (3): 310–318.

Atli, G.; Ariyurek, S. Y.; Kanak, E. G. and Canli, M. (2015). Alternations in the serum biomarkers belong to different metabolic systems of fish (*Oreochromis niloticus*) after Cd and Pb exposures. Toxicol. Pharmacol., 40 (2) :508-515.

AVMA (American Veterinary Medical Association). (2017). Judicious use of antimicrobials for treatment of aquatic animals by veterinarians.

Awasthi, K. K.; Awasthi, A.; Bhoot, N.; John, P.; Sharma, S. K. and Awasthi, K. (2013). Antimicrobial properties of electro-chemically stabilized organo-metallic thin films. Adv. Electrochem., 1 (1): 42–47.

Bakhshwan, S. A.; Marzouk, M. S.; Hanna, M. I. and Hamed, H. S. (2009). Some investigations on the clinical and biochemical alterations associated with dizinon toxicity in *Clarias gariepinus*. Egypt J. Aquat. Biol. Fish., 13: 173-197.

Bancroft, J. D.; Stevens, A. and Turner, D. R. (1996). Theory and practice of histological technique. 4th ed. Churchill, living stone and New York.

Barbieri, E. and Ferreira, L. A. A. (2011). Efffects of the organophosphate pesticide Folidol 600 on the freshwater fish, Nile tilapia (*Oreochromis niloticus*). Pestic. Biochem. Physiol., 99 (3): 209-214.

Bar-Ilan, O.; Albreecht, R. M.; Fako, V. E. and Furgeson, D. Y. (2009). Toxicity assessment of multisized gold and silver nanoparticles in zebrafish embryos. Small., 5 (16): 897-910.

Belitz, H. D.; Grosch, W. and Scieberle, P. (2001). Lehrbuch der lebensmittelchemie. Berlin: Springer 342 P. Res. J. A, Veterinary and Fishery sci., 2 (2): 17-23.

Benavides, M.; Fernández-lodeiro, J.; Coelho, P.; Lodeiro, C. and Diniz, M. S. (2016). Single and combined effect of aluminum (AL_2O_3) and zinc (ZnO) oxide nanoparticles in a freshwater fish

(Carassius auratus). Environ. Sci. Pollu. Res., 23 (24): 24578-24591.

Beutler, E.; Duron, O. and Kelly, B. M. (1963). Improved method for the determination of blood glutathione. J. Lab. Clin. Med., 61: 882-888.

Bruneau, A.; Turcotte, P.; Pilote, M.; Gagné, F. and Gagnon, C. (2016). Fate of silver nanoparticles in waste water and immunotoxic effects on rainbow trout. Aquat. Toxicol., 174: 70-81.

Brusle, J. and Anadon, G. G. I. (1996). The structure and function of fish liver, fish morphology. Horizon of New Research. Sci. Inc., 128.

Campos-Garcia, J.; Martinez, D. S. T.; Rezende, K. F. O.; Silva, J. R. M. C.; Alves, O. L. and Barbieri, E. (2016). Histopathological alterations in the gills of Nile tilapia exposed to carbofuran and multiwalled carbon nanotubes. Ecotoxicol. Environ. Saf., 133:481-488.

Chakeri, R.; Sajadi, M. M.; Kamrani, E. and Aghajari, N. (2015). Determination of heavy metal (lead and cadmium) concentrations in liver and muscle tissue of Indian mackerel (*Rastrelliiger kanagurta*) in Persian Gulf. Iran. Sci. Fish. J., 24 (2): 115-125.

Chelikani, P.; Fta, I. and Loewen, P.C. (2004). Diversity of structures and properties among catalases. Cell. Mol. Life Sci., 61 (2): 192-208.

Chen, C. Y. and Folt, C. L. (2000). Bioaccumulation and diminution of arsenic and lead in a freshwater food web. Environ Sci. Technol., 34: 3878- 3884.

Choi, T. E.; Kim, S.; Ahn, J. H.; Youn, P.; Kang, T. S.; Park, K.; Yi, J. and Ryu, D. Y. (2010). Induction of oxidative stress and apoptosis by silver nanoparticle in the liver of adult zebrafish. Aqua.Toxcol., 100: 151-159.

Cong, Y.; Banta, G. T.; Selck, H.; Berhanu, D.; Valsami-Jones, E. and Forbes, V. E. (2011). Toxic effects and bioaccumulation of nano- micron and ionic- Ag in the polychaete, Nereis diversicolor, 105: 403-411.

Craig, S.; Kuhn, D. D. and Schwarz, M. H. (2017). Understanding fish nutration, feeds and feeding. Virginia Cooperative Extension., 420: 256-266.

Dalibor, K.; Alzbeta, S.; Eliska, Z.; Antonin, K. and Josef, V. (2014). The effect of subchronic metribuzin exposure to signal crayfish (*Pacifastacus leniusculus)* Dana1852. Neuro. Endocrinol. Lett., 35 (2): 51–56.

Das, P. C.; Ayyappan, S. and Jena, J. K. (2004). Acute toxicity of ammonia and its sublethal effects on selected hematological and enzymatic parameters of *Cirrhinusmrigala mrigala* (Hamilton). Aquacult. Res., 35: 134-143.

Diab, A. S.; John, G.; Abd El Hady, Y. and Fathy, M. (2004). Evaluation of *Nigella sativa* (black seeds, barka), *Allium sativum* (garlic) & BIOGEN as a feed additive in fish culture *O. niloticus*. Scmj., 31 (2) :557-562.

Domingues, I.; Oliveira, R.; Lourenço, J.; Grisolia, C. K.; Mendo, S. and Soares, A. M. V. M. (2010). Biomarkers as a tool to assess effects of chromium (VI): comparison of responses in zebrafish early life stages and adults. Comp. Biochem. Physiol. Part (C): Toxicol. Pharmacol., 152 (3): 338-345.

Dos Santos, C. A.; Seckler, M. M.; Ingle, A. P.; Gupta, I.; Galdiero, S.; Galdiero, M.; Gade, A. and Rai, M. (2014). Silver nanoparticles: therapeutical uses, toxicity, and safety issues. J. Pharmaceutical. Sci., 103: 1931–1944.

Drake, P. L. and Hazelwood, K. J. (2005). Exposure related health effects of silver and silver compounds: A review. Ann. Occup. Hyg., 49 (7): 575-585.

Dumas, B. T. and Biggs, H. G. (1972). In stander method of clinical chemistry. Academic Press New York.

El-Bahr, S. M. and Azraqi, A. A. (2014). Effect of Dietary Supplementation of Turmeric (*Curcuma longa*) and Black Cumin Seed (*Nigella Sativa*) in Streptozotocin Induced Diabetic Rats. Inter. J. Biochm. Res. Rev., 4 (6): 481-492.

El–Far, A. H.; Korshom, M. A.; Mandour, A. A.; El-Bessoumy, A. A. and El-Sayed, Y. S. (2017). Hepatoprotective efficacy of *Nigella Sativa* seeds dietary supplementation against lead acetate-induced oxidative damage in rabbit – purification and characterization of glutathione peroxides. Biomed. Pharmacoth., 89: 711-718.

El-Gohary, R. A. L. and Marwa, A. M. G. (2011). Biochemical effect of three molluscicide baits against the two land snails, Monacha cantiana and Eobania vermiculata (Gastropoda: Helicidae). Int. J. Agric. Res., 6: 682-690.

Eminzade, S.; Uras, F. and Izzettin, F. V. (2008). Silymarin protects liver against toxic effects of anti-tuberculosis drugs in experimental animals. Nutr. Metab., 5: 18-25.

Ezzat, A. (2005). Effect of black seed (*Nigella sativa*) on the immune status of catfish (*Clarias gariepinus*) Ph. D Thesis, Fuculty of Vet. Med. Assiut Univ. Egypt.

Faheem, M.; Adeel, M.; Khaliq, S.; Lone, K. P. and Sayed, A. E. H. (2020). Bisphenol-A induced antioxidants imbalance and cytokines alteration leading to immune suppression during larva development of *Labeo rohita*. Environ. Sci. Pollut. Res., 27: 26800-26809.

Farmen, E.; Mikkelsen, H. N.; Evensen, O.; Einset, J.; Heier, L. S.; Rosseland, B. O.; Salbu, B.; Tollefsen, K. E. and Oughton, D. H. (2012). Acute and sub-lethal effects *in juvenile Atlantic* salmon exposed to low µg/L concentrations of Ag nanoparticles. Aquat. Toxicol., 108: 78–84.

Farooqui, Z.; Ahmed, F.; Rizwan, S.; Shahid, F.; Khan, A. A. and Khan, F. (2017). Protective effect of *Nigella sativa* oil on cisplatin induced nephrotoxicity and oxidative damage in rat kidney. Biomed. Pharmacother., 85: 7-15.

Federici, G.; Shaw, B. J. and Handy, R. D. (2007). Toxicity of titanium dioxide nanoparticles to rainbow trout (*Oncorhynchus mykiss*): gill injury, oxidative stress and other physiological effects. Aquat. Toxicol., 84 (4): 415-430.

Feng, Q. l.; Wu, J.; Chen, G. Q.; Cui, F. Z.; Kim, T. N. and Kim, J. O. (2000). A mechanistic study of the antibacterial effect of silver ions on Escherichia coli and staphylococcus aureus. J. Biomed. Mater. Res., 52 (4): 662-668.

Flower, N. A. L.; Brabu, B.; Revathy, M.; Gopalakrishnan, C.; Raja, S. V.; Murugan, S. S. and Kumaravel, T. S. (2012). Characterization of synthesized silver nanoparticles and assessment of its genotoxicity potentials using the alkaline comet assay. Mutat. Res., 18 (742): 61-65.

Fossati, P.; Prencipe, L. and Berti, G. (1980). Use of 3, 5-dichloro-2- hydroxybenzenesulfonic acid/ 4-aminophenazone chromogenic system in direct enzymic assay of uric acid in serum and urine. Clin. Chem., 26 (2) :227-231.

Foster, L. and Dunn, R. (1974). Single antibody technique for radioimmunoassay of cortisol in unextracted serum or plasma. Clin. Chem., 20: 365-372.

Frakas, J.; Christian, P.; Tollefsen, K. E.; Hylland, K. and Thomas, K. V. (2011). Uptake and effects of manufactured nanoparticles on rainbow trout (*Oncorhynchus mykiss*) gill cells. Aquat. Toxicol., 101:117-125.

Freeman, H. C.; Sangalang, G. B.; Uthe, J. F.; Garside, E. T. and Daye, P. G. (1983). A histopathological examination and analysis for polychlorinated hydrocarbons in inshore Atlantic cod (*Gadus morhae*). Arch. Environ. Contam. Toxicol. J., 12: 627-632.

Gagné, F.; Auclair, J.; Turcotte, P. and Gagnon, C. (2013). Sub lethal effects of silver nanoparticles and dissolved silver in freshwater mussels. J. Toxicol. Environ. Heal. Part A, 76: 479–490.

Gagné, F.; André, C.; Skirrow, R.; Gélinas, M.; Auclair, J.; Van, Aggelen, G.; Turcotte, P. and Gagnon, C. (2012). Toxicity of silver nanoparticles to rainbow trout: a toxicogenomic approach. Chemosphere, 89: 615-622.

Ghazala, M. S.; Al-Ghanim, K. A.; Sultana, S.; Al-Balawi, H. F. A.; Sultana, T.; Al-Misned, F. and Ahmed, Z. (2014). Acute toxicity II: effect of organophosphates and carbamates to *Catla catla* fingerlings. JAPS, 24: 1795-1801.

Ghosh, M.; Manivannan, J.; Sinha, S.; Chakraborty, A. and Mallick, S. K. (2012). *In vitro* and *in vivo* genotoxicty of silver nanoparticles. Mut. Res. Gen. Toxicol. Environ. Mutagen., 749 (1-2): 60-69.

Gonzalez, P.; Baudrimont, M.; Boudou, A. and Bourdineaud, J. P. (2006). Comparative effects of direct cadmium contamination on gene expression in gills, liver, skeletal muscles and brain of the zebra fish (*Danio rerio*). Biometals, 19 (3): 225-235.

Goreja, W. G. (2003). Black seed: nature's meracle remedy. Amazing Herbs Press, New York.

Govidasamy, R. and Abdul-Rahuman, A. (2012). Histopathological studies and oxidative stress of synthesized silver nanoparticles in Mozambique tilapia (*Oreochromis mossambicus*). J. Environ. Sci., 24 (6): 1091-1098.

Haggag, N. Z. (2004). Physiological studies on the changes induced some non- traditional pesticides to the Nile catfish. MSc. D. Thesis. Faculty of Women, Ain Shams Univ. Cairo, Egypt.

Hamed, H. S. and Abdel–Tawwab, M. (2017). Ameliorative effect of propolis supplementation on alleviating bisphenol-A toxicity: Growth performance, biochemical variables, and oxidative stress biomarkers of Nile tilapia *(Oreochromis niloticus)* (L.) fingerlings. Comp. Biochem. Physiol. (C) Toxicol. Pharmacol., 202: 63-69.

Hamed, H. S. and Abdel–Tawwab, M. (2021). Dietary pomegranate (*Punica granatum*) peel mitigated the adverse effects of silver nanoparticles on the performance, haemato-biochemical, antioxidant, and immune responses of Nile tilapia fingerlings. Aquaculutre, 540: 736-742.

Hamed, H. S. and Osman, A. G. M. (2017). Modulatory effect of lycopene against carbofuran toxicity in African catfish (*Clarias gariepinus)*. Fish. Physiol. Biochem., 43:1721-1731.

Hamed, H. S.; Ismal, S. M. and Faggio, C. (2021). Effect of allicn on antioxidant defense system, and immune response after carbofuran exposure in Nile tilapia (*Oreochromis niloticus*). Comp. Biochem. Physiol. Part (C) Toxicol. Pharmacol. (240): 108919.

Hasan, M. A.; Mustapha, N. M.; Kadir, A. A. and Hezmee, M. (2018). Potential role of *Nigella sativa* (NS) in abating oxidative stress. induced toxicity in rats: a possible protection mechanism. IOSR. J. Pharm. Biolo. Sci., 13: 29-42.

Hedayati, A.; Kolangi, H.; Jahanbakhshi, A. and Shaluei, F. (2012). Evaluation of silver nanoparticles ecotoxicity in silver carp (*Hypophthalmicthys molitrix*) and goldfish (*Carassius auratus*). Bulga. J. Vet. Med., 15 (3): 172-177.

Hedayati, S. A.; Farsani, H. G.; Naserabad, S. S.; Hoseinifar, A. H. and Doan, H. V. (2019). Protective effect of dietary vitamin E on immunological and biochemical induction through silver nanoparticles (Ag-NPs) inclusion in diet and silver salt (AgNO3) exposure on zebrafish (*Dnio rerio*) Comparative Biochemistry and Physiolog part C: Toxicol. Pharmacol., 222: 100-107.

Hossain, Z. and Hug, F. (2002). Studies on the interaction between Ag^{+} and DNA. J. Inorg. Biochem., 91: 398-404.

Hunag, C. H.; Chang, S. L. and Hung, W. (2003). Dietary vitamin E supplementation affects tissue lipid peroxidation of hybrid tilapia (*Oreochromis niloticus*) x. aureus. Comp. Biochem. Physiol. B: Biochem. Mol. Biol., 134 (2): 265-270.

Ijaz, H.; Tulain, U. R.; Qureshi, J.; Danish, Z.; Musayab, S.; Akhtar, M. F.; Saleem, A.; Khan, K. K.; Zaman, M.; Waheed,

I.; Khan, I. and Abdel-Daim, M. (2017). Review: *Nigella sativa* (Prophetic Medicine): A review. Pak. J. Pharm. Sci., 30: 229- 234.

Imani, M.; Halimi, M. and Khara, H. (2015). Effects of siver nanoparticles (Ag-NPs) on hematological parameters of rainbow trout *(Oncorhynchus mykiss)*. Comp. Clin. Pathol., 24 (3): 491-495.

Jayaseelan, C.; Abdul Rahuman, A.; Ramkumar, R.; Perumal, P.; Rajakumar, G.; Kirthi, A. V.; Kumar, T. S. and Marimuthu, S. (2014). Effect of subacute exposure to nickel nanoparticles on oxidative stress and histopathological changes in Mozambique tilapia *(Oreochromis mossambicus)*. Ecotoxicol. Environ. Saf., 107: 220-228.

Jin, Y. X.; Zhang, X. X.; Shu, L. J.; Chen, L. F.; Sun, L.W.; Qian, H.; Liu, W. and Fu, Z. (2010). Oxidative stress response and gene expression with atrazine exposure in adult female zebra fish (*Danio rerio*). Chemosphere, 78: 846-852.

Johari, S. A.; Kalbassi, M. R.; Soltani, M. and Yu, I. J. (2013). Toxicity of comparison of colloidal silver nanoparticles in various life stages of rainbow trout (*Oncorhynchus mykiss*). Iran. Fish. Sci., 12 (1): 76-95.

Johari, S. A.; Kalbassi, M. R.; Yu, I. J. and Lee, J. H. (2015). Cronic effect of waterborne silver nanoparticles on rainbow trout (*Oncorhynchus mykiss*): Histopathology and bioaccumulation. Comp. Clin. Pathol., 24 (5): 995-1007.

Kadry, S. M.; Marzouk, M. S.; Amer, A.F.; Hanna, M. I.; Azmy, A. H. and Hamed, H. S. (2012). Vitamin E as antioxidant in female African catfish *(Clarias gariepinus)* exposed to chronic toxicity of atrazine. Egypt. J. Aquat. Biol. Fish., 16: 83-98.

Kaegi, R.; Sinnet, B.; Zuleeg, S.; Hagendorfer, H.; Mueller, E.; Vonbank, R. and Burkhardt, M. (2010). Release of silver nanoparticles from outdoor facades, Environ. Pollu., 158: 2900-2905.

Kanter, m.; Coskun, O. and Gurel, A. (2005). Effect of black cumin (*Nigella sativa*) on cadmium-induced oxidative stress in the blood of rats. Biol. Trace. Elem. Res., 107: 277-287.

Kanter, M.; Coskun, O.; Kalayci, M.; Buyukbas, S. and Cagav, F. (2006). Neuroprotective effects of *Nigella sativa* on experimental spinalcord injury in rats. Hum. Exp. Toxicol., 25: 127–133.

Karami, A.; Christianus, A.; Ishak, Z.; Courtenay, S. C.; Sayed, M. A.; Noor, A. M. and Noor, S. H. (2010). Effect of triploidization on juvenile African catfish (*Clarias gariepinus*). Aquacult. Inter., 18: 851-858.

Kavitha, P. and Rao, J. V. (2007). Oxidative and locomotor behavior response as biomarkers for assessing recovery status of mosquito fish Gambusia affinis after lethal effect of an organo-phosphate pesticide, monocrotophos. Pest. Biochem. Physiol., 87: 182-188.

Kavitha, C.; Malarvizhi, A.; Senthil, K. S. and Ramesh, M. (2010). Toxicological effects of arsenate exposure on hematological, biochemical and liver transaminases activity in an Indian major carp *(Catla catla)*. Food Chem. Toxicol., 48: 2848-2854.

Kaya, H. and Akbulut, M. (2015). Effect of waterborne lead exposure in Mozambique tilapia: oxidative stress, osmoregulatory responses, and tissue accumulation. J. Aquat. Anim. Health., 27: 77-87.

Kim, S. H.; Ko, J. W.; Koh, S. K.; Lee, I. C.; Son, J. M.; Moon, C. and Kim, J. C. (2014). Silver nanoparticles induce apoptotic cell death in cultured cerebral cortical neurons, molecular& cellular. Toxicology, 10: 173-179.

Koracevic, D.; Koracevic, G.; Djordjevic, V.; Andrejevic, S. and Cosic, V. (2001). Method for the measurement of antioxidant activity in human fluids. J. Clin. Pathol., 54: 356-361.

Kori-Siakpere, O.; Ikomi, R. B. and Ogbe, M. G. (2012). Variations in acid phosphatase and alkaline phosphatase activities in the plasma of the African catfish (*Clarias gariepinus*) exposed to sublethal concentrations of potassium permanganate. Asian. J. Exp. Biol. Sci., 1 (1): 170-174.

Kotb, A. M.; Abdel El-Kareem, M.; Abou Khalil, N. S. and Sayed, A. E. H. (2018). Protective effect of *Nigella* on 4-nonylphenol induced nephrotoxicity in *Clarias gariepinus* (Burchell, 1822). Sci. Environ., 27: 619-620.

Kourouma, A.; Keita, H.; Duan, P.; Quan, C.; Bilivogui, K. K.; Qi, S. and Yand, K. (2015). Effect of 4-nonylphenol on oxidant/ antioxidant balance system inducing hepatic steatosis in male rat. Toxicol. Rep., 2: 1423-1433.

Kovacik, A.; Tvrda, E. and Miskeje, M. (2019). Trace metal in the freshwater fish *Cyprinus carpio*: Effect to serum biochemistry and oxidative status markers. Biol. Trace. Elem. Res., 188: 494-507.

Kumar, N.; Krishnani, K. K.; Gupta, S. K. and Singh, N. P. (2018). Effect of silver nanoparticles on stress biomarkers of Channa striatus: immuno-protective or toxic. Environ. Sci. Pollut. Res., 25: 14813-14826.

Latif, M.; Faheem, M.; Hoseinifar, S. H. and Doan, H. V. (2021). Dietary black seed effects on growth performance, proximate composition, antioxidant and histo-biochemical parameters of a culturable fish, rohu (*Labeo rohita*). Animals, 27: 11- 48.

Lee, B.; Duong, C. N.; Cho, J.; Lee, J.; Kim, K.; Seo, Y.; Kim, P.; Choi, K. and Yoon, J. (2012). Toxicity of citrate-capped silver nanoparticles in common carp (*Cyprinus carpio*). J. Biomed. Biotechnol. Article., 262670: 1–14.

Leong, X. F.; Mustafa, M. R. and Jaarin, K. (2013). *Nigella sativa* and its protective role in oxidative stress and hypertension, Review Article. Evid. Based Complement. Alternat. Med., 28: 1-9.

Li, W., Yin, D., Zhou, Y., Hu, S. and Wang, L.; (2003). 3, 4-Dichloroaniline- induced oxidative stress in liver of crucian carp (*Carassius auratus*). Ecotoxicol. Environ. Saf. 56:251-255.

Liu, Y.; Sun, L.; Yang, G. and Yang, Z. (2020). Nephrotoxicity and genotoxicity of silver nanoparticles in juvenile rats and possible mechanisms of action. Arh. Hig. Toksikol., 71: 121-129.

Litchfield, J. T. J. and Wilcoxon, F. (1949). A simplified method of evaluating dose-effect experiments. J. Pharmacol. Exp. Ther., 96 (2): 99-113.

Luther, M.; Parry, J.; Moore, J.; Meng, J.; Zhang, Y.; Cheng, Z. and Yu, l. (2007). Inhibitory effect of Chardonnay and black raspberry seed extracts on lipid oxidation in fish oil and their radical scavenging and antimicrobial properties. Food Chem., 104 (3): 1065-1073.

Maduenho, L. P. and Martinez, C. B. (2008). Acute effects of diflubenzuron on the freshwater fish Prochilodus lineatus. Comp. Biochem. Physiol. Part (C), 148: 265-272.

Mahmoud, U. M.; Mekkawy, I. A. A.; Naguib, M. and Sayed, A. E. H. (2019). Silver nanoparticle- induced nephrotoxicity in *Clarias gariepinus*: physio-histological biomarkers. Fish Physiol. Biochem., 45 (6): 1895-1905.

Mahmoud, H. S.; Almallah, A. A.; El-Hak, H. N. G.; Aldayel, T. S.; Abdelrazek, H. M. A. and Khaled, H. E. (2021). The effect of dietry supplementation with *Nigella sativa* (black seeds) mediates immunological function in male wistar rats. Sci. Rep., 11: 7542-7560.

Maita, M.; Shiomitsu, K. and Ikeda, Y. (1984). Health assessment by the climogram of hemochemical constituents in cultured yellow tail. Bull. Japn. Soc. Sci. Fish., 51: 205-211.

Manjula, D.; Dana, A. T.; John, T. B. and Milton, F. (1996). Accumulation, physiological and biochemical effect of cadmium in a simple aquatic food chain. Ecotoxicol. Environ. Saf., 33 (1): 38-43.

Marie, M. A.; Mohamed, H. A. and El–Badawy, A. A. (1998). Physiological and biochemical responses of the common carp *(Cyprinus carpio)* to an organophosphorous insecticide "profenofos". Egypt. J. Zool., 31: 279-302.

Marzouk, M. S.; Saleh, A. A.; Hammed, H. S. and Abdel-Daim, S. M. (2017). Assessment of physiological and antioxidant responses in African catfish (*Clarias gareipinus)* as biomarkers of metal pollution in River Nile. Wulfenia, 24 (10): 247-271.

Massarsky, A.; Abraham, R.; Nguyen, K. C.; Rippstein, P.; Tayabali, A. F.; Trudeau, V. L. and Moon, T. W. (2014). Nanosilver cytotoxicity in rainbow trout (*Oncorhynchus mykiss*) erythrocytes and hepatocytes. Comparative Biochemistry and Physiology Part C: Pharmacol.Toxicol. Endocrinol., 159: 1-21.

Maxine, M. and Benjamine, B. S. (1985). Outline of veterinary clinical pathology 3 rd ed. Colorado State Univ. USA.

Mekkawy, I. A.; Mahmoud, U. M. and Sayed, A. H. (2011). Effects of 4-nonylphenol on blood cells of the African catfish *(Clarias gariepinus)* (Burchell, 1822). Tissue and cell, 43: 223-229.

Mekkawy, I. A.; Hussein, S. Y.; Abdel-Nasser, M. and Ahmed, S. M. (1996). Comparative studies on the effect of herbicide atrazine on some blood constituents and protein electrophoretic patterns of *Oreochromis niloticus* and *chrysichthyes auratus*. Assuit. J. Egypt. Ger. Soc. Zool., 19 (A): 283-319.

Mohammed, T. A. (2003). Study of growth and reproduction of some corals at Hurghada region, with references to the effect of some pollutants in the area. Ph. D. thesis, Zoology Department, Faculty of Science, Suez Canal University, 204 pp.

Mohammed, H. H. and Arias, C. R. (2016). Protective efficacy of *Nigella sativa* seeds and oil against columnaris disease in fishes. J. Fish. Dis., 39: 693-703.

Mommsen, T. P.; Vijayan, M. M. and Moon, T. W. (1999). Cortisol in teleosts: dynamics, mechanisms of action, and metabolic regulation. Rev. Fish. Biol. Fish., 9: 211-268.

Monfared, A. L. and Soltani, S. (2013). Effects of silver nanoparticles administration on the liver of rainbow trout (*Oncorhynchus mykiss*): Histological and biochemical studies. Eur. J. Exp. Biol., 3 (2): 285-289.

Moraes, B. S.; Loro, V. L.; Pretto, A.; da Fonseca, M.B.; Menezes, C.; Marchesan, E.; Reimche, G.B. and de Avila, L.

A. (2009). Toxicological and metabolic parameters of the teleost fish, *Leporinus obtusidens* in response to commercial herbicides containing clomazone and propanil. Pest. Biochem. Physiol., 95: 57-62.

Morones, J. R.; Elechiguerra, J. L.; Camacho, A.; Holt, J. B.; Ramirez, J. T. and Yacaman, M. J. (2005). The bactericidal effect of silver nanoparticles. J. Nanotechnology, 16 (10): 2346-2353.

Muhammad, A. R. and Al-Ghamdi, M. S. (2002). A review of the pharmaco-therapeutic effects of *Nigella sativa*. Pak. J. Med. Res., 41(2): 77-83.

Naguib, M.; Mohamad, U. M.; Mekkawy, I. A. and Sayed, A. E. D. (2020). Hepatotoxic effect of silver nanoparticles on *Claries gariepinus*; Biochemical, histopathological, and histochemical studies. Toxicol. Rep., 7: 133-141.

Nishikimi, M.; Appaji, N. and Yagi, K. (1972). The occurrence of superoxide anion in the reaction of reduced phenazine methosulfate and molecular oxygen. Biochem. Biophys. Res. Commun., 46: 849-854.

Nowack, B.; Krug, H. F. and Height, M. (2011). 120 years of nanosilver history: implications for policy makers. Environ. Sci.Technol., 45 (4): 1177–1183.

NRC, (2012). A research Strategy for Environmental, Health, and Safety Aspects of Engineered Nanomaterials. The National Academies Press, Washington, DC.

Ohkawa, H.; Ohishi, N. and Yagi, K. (1979). Assay for lipid peroxide in animal tissues by thiobarbituric acid reaction. Anal. Biochem., 95 (2): 351-358.

Öner, M.; Atli, G. and Canli, M. (2008). Changes in serum biochemical parameters of freshwater fish *Oreochromis niloticus* following prolonged metal (Ag, Cd, Cr, Cu and Zn) exposure. Environ. Toxicol. Chemist., 27 (2): 360-366.

Orsonneau, J. L.; Douet, P.; Massoube, C.; Lustenberger, P. and Bernard, S. (1989). An improved pyrogallol red-molybdate method for determining Total Urinary Protein. Clin. Chem., 35: 2233-2236.

Ostaszewska, T.; Śliwiński, J.; Kamaszewski, M.; Sysa, P. and Chojnacki, M. (2018). Cytotoxicity of silver and copper nanoparticles on rainbow trout (*Oncorhynchus mykiss*) hepatocytes. Environ. Sci. Pollut. Res. Int., 25 (1): 908-915.

Pandit, D. N. and Sinha, A. (2019). Ecotoxicity of silver nanoparticles in an Indian air-breathing catfish *(Clarias batrachus)* (*Linnaeus*). J. Entomol. Zool. stud., 7 (5): 906-911.

Park, S. and Choi, J. (2010). Geno-and ecotoxicity evaluation of silver nanoparticles in freshwater *(crustacean Daphnia magna).* Environ. Eng. Res., 15 (1): 23-27.

Pournori, B.; Paykan Heyrati, F. and Dorafshan, S. (2017). Histological changes in various tissues of striped catfish, pangasianodon hypophthalmus, fed on dietary nucleotides and exposed to water-borne silver nanoparticles or silver nitrate. Iran. J. Aquat. Anim. Health., 3: 36-52.

Prasanth, M. S.; David, M. and Mathed, S. G. (2005). Behavioral changes in freshwater fish *(Cirrrhinus mrigala Hamilton*) exposed to cypermethrin. J. Ecotoxicol. Environ. Monit., 26: 141-144.

Qiu, W.; Yang, M.; Liu, J.; Xu, H.; Luo, S.; Wong, M. and Zheng, C. (2019). Bisphenol S-induced chronic inflammatory stress in liver via peroxisome proliferator-activated receptor Ɣ using fish *in vivo* and *in vitro* models. Environ. Pollut., 246: 963-971.

Radwan, O. A. and El-Said, M. M. (2006). Biochemical studies on residues of two different formulations of profenofos insecticidein *Oreochromis niloticus*. J. Biol. Chem. Environ. Sci., 1(3): 491-519.

Rajkumar, K. S.; Kanipandian, N. and Thirumurugan, R. (2015). Toxicity assessment on haemotology, biochemical and histopathological alteration of silver nanoparticles-exposed freshwater fish *Labeo rohita*. Appl. Nanosci., 6 (1): 19-29.

Rathore, R. S. and Khangarot, B. S. (2002). Effect of temperature on the sensitivity of sludge worm Tubifex tubifex (Muller) to selected heavy metals. Ecotoxicol. Environ. Saf., 53: 27-36.

Reddy, P. K. and Leatherland, J. F. (1998). Stress physiology, in: fish disease and disorders, Vol.2 Non-infection Disorders. (Eds. Leatherland, J. F. & Woo, P. T. K.). Cabi. Pub., 279-301.

Regoli, F.; Gorbi, S.; Fattorini, D.; Tedesco, S.; Notti, A.; Machella, N.; Bocchetti, R.; Beneddetti, M. and Piva, F. (2006). Use of the land snail Helix aspersa as sentinel organism for monitoring ecotoxicologic effects of urban pollution: an integrated approach. Environ. Health. Perspectives, 114: 63-69.

Ribeiro, F.; Gallego-Urrea, J. A.; Jurkschat, K.; Crossley, A.; Hassellöv, M.; Taylor, C.; Soares, A. M. V. and Loureiro, S. (2014). Silver nanoparticles and silver nitrate induce high toxicity to Pseudokirchneriella subcapitata: *Daphnia magna Daphnia magna* and *Danio rerio*. Sci. Total Environ., 466–467, 232–241.

Roberts, R. J. (2001)."Fish pathology ". 3rd. Bailliere Tindall, London, England.

Rothkamm, K. and Lobrich, M. (2003). Evidence for a lack of DNA double-strand break repair in human cells exposed to very low X-ray dose. Proc. Natl. Acad. Sci., U.S.A. 100: 5057-5062.

Sakai, T.; Murata, H.; Yamauchi, K.; Sekiya, T. and Ukawa, M. (1992). Effects of dietry lipid peroxides contents on *in vitro* lipid peroxidation, a- tochopherol contents and superoxide dismutase and glutathione peroxidase activities in the liver of yellow tilapia. Nippon Suisan Gakkaishi, 58: 1483-1486.

Salah, M. A.; Mona, M. I.; Mohamed, F. and Mohamed, A. A. (2019). The role of *nigella sativa* in improving the immune response of the African catfish (*Clarias gariepinus*) to *Aeromonas hydrophila* vaccine. Egypt. J. Aquat. Biol. Fish., 23 (4) :373-384.

Samipillai, S. and Jagadeesan, G. (2005). Influence of taurine on phosphatases activity in selected tissues of mercuric chloride intoxicated mice. J. Exp. Zool., 8: 295-300.

Sayed, A. H. (2016). Genotoxicity detection following exposure to silver nanoparticles in African catfish (*Clarias gariepinus*). Int. J. Nanopart., 9 (1): 41–53.

Sayed, A. E. and Hamed, H. S. (2017). Induction of apoptosis and DNA damage by 4-nonylphenol in African catfish (*Clarias gariepinus*) and the antioxidant role of Cydonia oblonga. Ecotoxicol. Environ. Saf., 139- 97:101.

Sayed, A. E. and Soliman, H. A. M. (2017). Developmental toxicity and DNA damaging properties of silver nanoparticles in the catfish (*Claries gariepinus*). Mutation.Res./ Genetic. Toxicol. Environ. Mutagen., 822: 34-40.

Sayed, A. H. and Soliman, H. A. M. (2018). Modulatory effects of green tea extract against the hepatotoxic effects of 4-nonylpenol in catfish *(Clarias gariepinus)*. Eco. Toxicol. Environ. Saf., 149: 159-165.

Sayed, A. E.; Mohamed, N. H.; Ismaeil, M. A.; Abdel-Mageed, W. M. and Shoreit, A. A. (2016). Antioxidant and antiapoptotic activities of calotropis procera latex on catfish (*Clarias gariepinus)* exposed to toxic 4- nonylphenol. Ecotoxicol. Environ. Saf., 128: 189-194.

Scheletter, G. and Nussel, E. (1975). Arbeitsmed sozialmed pracentimed, 28: 10-25.

Scown, T. M.; Santos, E. M.; Johnston, B. D.; Gaiser, B.; Baalousha, M.; Mitov, S.; Lead, J. R.; Stone, V.; Fernandes, T. F.; Jepson, M.; van Aerle, R. and Tyler, C. R. (2010). Effects of aqueous exposure to silver nanoparticles of different sizes in Rainbow Trout. Toxicol. Sci., 115 (2): 521-534.

Shahid, F.; Farooqui, Z.; Rizwan, S.; Abidi, S.; Parwez, I. and Khan, F. (2017). Oral administration of *Nigella sativa* oil ameliorates the effect of cisplatin on brush border membrane enzymes, carbohydrate metabolism and antioxidant system in rat intestine. Exp. Toxicol. Pathol., 69: 299–306.

Shalaby, A. M. (2001). Protective effect of ascorbic acid against mercury intoxication in Nile tilapia (*Oreochromis niloticus*). J. Egypt, Acad. Soc. Environ. Dev., 2: 79-97.

Shaluei, F.; Hedayati, A.; Jahanbakhshi, A.; Kolangi, H. and Fotovat, M. (2013). Effect of subacute exposure to silver nanoparticle on some hematological and plasma biochemical indices in silver carp (*Hypophthmichthys molitrix*). Hum. Exp. Toxicol., 32: 1270-1277.

Singh, A.; Dar, M. Y.; JoshI, B.; Sharma, B.; Shrivastava, S. and Shukla, S. (2018). Phytofabrication of silver nanoparticles: novel drug to overcome hepatocellular ailments. Toxicol. Rep., 5: 333-342.

Slatinská, I.; Smutná, M.; Havelková, M. and Svobodová, Z. (2008). Review article: Biochemical markers of aquatic pollution in fish- Glutathione-S-transferase. Folia. Vet., 52: 129-134.

SPSS, (2004). "Statistical and package for social science, SPSS for windows release 14.0.0, 19June, 2004." Standard version, copyright SPSS Lnc., 1989-2004, All rights Reserved, copyright 2 SPSS Lnc.

Stein, E. A. (1986). In textbook of clinical chemistry, W. B. Saunder, Philadelphia, 28: 879-886.

Stensberg, M. C.; Wei, Q.; McLamore, E. S.; Porterfield, D. M.; Wei, A. M. S. and Sepúlveda, M. S. (2011). Toxicological studies on silver nanoparticles: challenges and opportunities in assessment, monitoring and imaging, Nanomedicine, 6: 879-898.

Stružyński, W.; Dąbrowska-Bouta, B.; Grygorowicz, T., Ziemińska, E. and Stružyńska, L. (2013). Markers of oxidative stress in hepatopancreas of crayfish (*Orconectes limosus*, raf) experimentally exposed to nanosilver. Environ. Toxicol., 29 (11): 1283-1291.

Taju, G.; Majeed, S. A.; Nambi, K. and Hameed, A. S. (2014). *In vitro* assay for the toxicity of silver nanoparticles using heart and gill cell lines of *Catla catla* and gill cell line of *Labeo rohita*. Comp. Biochem. Physiol. C Pharmacol. Toxicol., 161: 41–52.

Taşar, N.; Şehirli, O.; Yigner, O., Süleymanoğlu, S.; Meral, Y. and Yeğen, Ş. G. (2012). Protective effects of *Nigella sativa* against hypertension-induced oxidative stress and cardiovascular dysfunction in rats. Marmara. Pharm. J.,16 (2) :141–149.

Terzi, A.; Coban, S.; Yildiz, F.; Ates, M.; Bitiren, M.; Taskin, A. and Aksoy, N. (2010). Protective effects of *Nigella sativa* on intestinal ischemia-reperfusion injury in rats. J. Invest. Surg., 23 (1): 21–27.

Tietz, N. W. (1986). Textbook of Clinical Chemistry, W. B. Saunders Company, Philadelphia, 28: 1270- 1271.

Tietz, N. W. (1994). Textbook of Clinical Chemistry, 2 nd ed. W. B. Saundes Co., Philadelphia, 831-832.

Tietz and Norbert, W. (1995). Clinical Guide to Laboratory Tests, 3 rd Edition, WB Saunders, Philadelphia, 186-188.

Üral, M. S. and Koprucu, S. S. (2006). Acute toxicity of dichlorvos on fingerling European catfish (*Silurus glanis L*). Bull Environ. Contam. Toxicol., 76 (5): 871-876.

Vajargah, M. F.; Imanpoor, M. R.; Shabani, A. and Hedayati, A. (2019). Effect of long-term exposure of silver nanoparticles on growth indices, hematological and biochemical parameters and gonad histology of male goldfish (*Crassius auratus gibelio*). Microsc. Res. Tech., 82 (7): 1224-1230.

Valavanidis, A.; Vlahogianni, T.; Dassenakis, M. and Scoullos, M. (2006). Molecular biomarkers of oxidative stress in aquatic organisms in relation to toxic environmental pollutants. Ecotoxicol. Environ. Saf., 64: 178-189.

Valerio-Garcíaa, R. C.; Carbajal-Hernándeza, A. L.; Martínez-Ruíza, E. B.; Jaruín-Diaza, V. H.; Haro-Pērezb, C. and Martínez-Jerónimo, F. (2017). Exposure to silver nanoparticles produces oxidative stress and affects macromolecular and metabolic biomarker in the goodeid fish *(Chapalichthys pardalis)*. Sci. Total. Environ., 1(583): 308-318.

Vassault, A. (1983). Lactate dehydrogenase: UV-method with pyruvate and NADH. In Methods of enzymatic analysis, 3rd ed., Bergmeyer HU., Ed.; Plenum: New York, III, pp. 118-125.

Velisek, J.; Stara, A.; Kolarova, J. and Svobodova, Z. (2011). Biochemical, physiological and morphological responses in common carp, *(Cyprinus carpio L)*. after long term exposure to terbutryn in real environmental concentration. Pest. Biochem. Physiol., 100: 305-313.

Vignesh, V.; Anbarasi, K. F.; Karthikeyeni, S.; Sathiyanarayanan, G.; Subramanian, P. and Thirumurugan, R. (2013). A superficial phyto-assisted synthesis of silver nanoparticles and their assessment on hematological and biochemical parameters in *Labeo rohita* (Hamilton, 1882), Colloids Surf., A: Physicochem. Eng., 439: 184-192.

Wang, B.; Feng, W. Y.; Wang, M.; Wang, T. C.; Gu, Y. Q.; Zhu, M. T.; Ouyang, H.; Sho, J. W.; Zhang, F.; Zhao, Y. L.; Chai, Z. F.; Wang, H. F. and Wang, J. (2008). Acute toxicological impact of nano- and submicro-scaled zinc oxide powder on healthy adult mice, J. Nanopart. Res., 10: 263-276.

Wijnhoven, S. W.; Peijnenburg, W. J. G.; Herberts, C. A.; Hagens, W. I.; Oomen, A. G.; Heugens, E. H. W.; Roszek, B.; Bisschops, J.; Ilse, G.; De, M. D.V.; Dekkers, S.; De, J. W. H.; van, Z. M.; Sips, A. J. M. and Geertsma, R. E. (2009). Nano-silver a review of available data and knowledge gaps in human and environmental risk assessment. Nanotoxicology, 3 (2):109–138.

Woodrow Wilson International Center for Scholars (2008). The project on emerging nanotechnologies. Consumer Products. An Inventory of Nanotechnology-Based Consumer Products Currently on the Market [Cited 2008 May 31] (http://www.nanotechproject. org/inventories/consumer/).

Wu, Y. and Zhou, Q. (2013). Silver nanoparticles cause oxidative damage and histological changes in medaka (*Oryzias latipes*) after 14 days of exposure. Environ. Toxicol. Chem., 32(1)165-173.

Wu, Y. Q.; Zhou, H.; Liu, T. and Wang, G. J. (2010). Effects of silver nanoparticles on the development and histopathology biomarkers of Japanese medaka (*Oryzias latipes*) using the partial-life test. Aquat. Toxicol., 100: 160-167.

Xiong, D.; Fang, T.; Yu, L.; Sima, X. and Zhu, W. (2011). Effect of nano-scale TiO_2, ZnO and their bulk counterparts on zebrafish: Acute toxicity, oxidative stress and oxidative damage. Scien.Total. Environ., 409 (8): 1444-1452.

Yalsuyi, A. M and Vajargah, M. F. (2017). Acute toxicity of silver nanoparticles in Roach (*Rutilus rutilus*) and Goldfish (*Carassius auratus*) J. Environ. Treat., Tech. 5(1):1-4.

Yang, J. L. and Chen, H. C. (2003). Effect of gallium on common carp (*Cyprinus carpio*): acute test, serum biochemistry, and erythrocyte morphology. Chemosphere, 53 (8): 877-882.

Yazdanparast, T.; Sharifpour, I.; Soltani, M. and Esfahani, H. K. (2016). Evaluation of silver retention in different organs of zebrafish (*Danio Rerio*) fed diet supplemented with silver nanoparticles. Int. J. Eng. Res., 5: 269-274.

Yokozawa, T. and Dong, E. (2001). Role of ginsenoside-Rd in cisplatin-induced renal injury: special reference to DNA fragmentation Nephron., 9: 433-438.

Young, D. S. (1990). Effects of drugs on clinical laboratory tests, 3 rd Ed., 3: 168-182.

Yousafzai, A. M. and Shakoori, A. R. (2011). Hepatic responses of a freshwater fish against aquatic pollution. Pakistan J. Zool., 43 (2): 209-221.

Yuan, C.; Zhang, Y.; Liu, Y.; Zhang, T. and Wang, Z. (2016). Enhanced GSH synthesis by Bisphenol A Exposure Promoted DNA methylation process in the Testes of Adult rare minnow Gobiocypris rarus. Aqua.Toxicol., 178: 99-105.

Zebral, Y. D.; Fonseca, J. D. S.; Costa, P. G.; Oliveira, C. S. D.; Zocke, T. G.; Pizzol, J. L. D.; Robaldo, R. B. and Bianchini, A. (2019). Waterborne copper is more toxic to the killifish *(Poecilia vivipara)* in elevated temperatures: linking oxidative stress in the liver with reduced organismal thermal performance. Aquat. toxicol., 209: 142-149.

Zhang, J. F.; Shen, H.; Wang, X. R.; Wu, J. C. and Xue, Y. Q. (2004). Effects of chronic exposure of 2, 4 dichlorophenol on the antioxidant system in liver of freshwater fish *(Carassius auratus)*. Chemosphere, 55: 167-174.

الملخص العربى

تأثير حبة البركة على جزيئات الفضة النانوميترية المحدثة للتسمم الكبدي والكلوي فى أسماك القرموط

تم التخطيط للعمل الحالي لدراسة تأثير جزيئات الفضة النانوميترية علي معايير كيميائية حيوية مختارة للمصل وأنسجة الكبد والكلي فى أسماك القرموط الأفريقي. أجريت هذه الدراسة أيضا لتقييم تأثير جزيئات الفضة النانوميترية على الإجهاد التأكسدية والإنزيمات المضادة للأكسدة وتدمير الحمض النووي. بالإضافة إلي ذلك هدفت الدراسة إلي تصليت الضوء علي التغييرات النسيجية المحدثة من التعرض لجزيئات الفضة النانوميترية . وإظهار دور الحبة السوداء في تحسين هذه التغيرات.

التصميم التجريبي والمعاملات التى تم فحصها:

لهذه الدراسة الحالية تم أستخدام ١٦٨ سمكة من أسماك القرموط الأفريقي أخذت ٤٨ سمكة لتعيين نصف التركيز المميت لجزيئات الفضة النانوميترية. تم تقسيم ١٢٠ سمكة الأخري إلى أربع مجموعات. كل مجموعة تتكون من٣ متكررات بحيث يكون كل ١٠ أسماك لكل حوض مائي. المجموعة الأولى كانت بمثابة المجموعة الضابطة. وعملت المجموعة الثانية كمجموعة تحكم وتم إضافه ٣% من حبة البركة. تعرضت المجموعة الثالثة الى ٥٠ ملجم/ لتر من جزيئات الفضة النانوميترية. تم تعرض المجموعة الرابعة إلى ٥٠ ملجم/ لتر من جزيئات الفضة النانوميترية بالإضافه إلى ٣% من حبة البركة.

بعد ٣٠ يوماً من المعاملة، تم جمع عينات الدم من الوريد. تم جمع مصل الدم واستئصلت أنسجة الكبد والكلي بسرعة من أجل الدراسات البيوكيميائية والنسيجية.

تضمن العمل الحالي الفحوصات التالية:

- الفحص المرضي وما بعد الموت.
- المصل: تحديد (الفوسفاتيز القلوي، الألانين أمينو ترانسفيريزو أسبارتات أمينو ترانسفيراز) بالأضافه إلى وظائف الكلي (اليوريا & الكرياتينين و حمض اليوريك) والبروتين ومستوى السكر فى الدم والكورتيزول وإنزيم نازع لهيدروجين اللاكتات بالدم

- ومستوى الدهون بالدم .
- تحليل تغيرات الحمض النووي
- الفحص النسيجي للكبد والكلي

النتائج:

تم ملاحظة نصف التركيز المميت لجزيئات الفضة الناوميترية كان ١٧١,٧ مجم/ لتر فى ٩٦ ساعة.

المعايير البيوكيميائية:

تتجلى العلامات المرضية لسمك القرموط الأفريقي المرتبطة بسمية ٥٠ مجم/ لتر من الفضة النانوميترية خلال أعراض عصبية بما فى ذلك سلوك السباحة غير الطبيعي في شكل غير منتظم ، وتصبغ الجلد غير طبيعي على شكل ذبول يزيد من إفراز المخاط والحركات السريعة. تم الكشف عن نتائج التشريح بدرجات متفاوتة خاصة فى جميع الأعضاء الداخلية لأسماك القرموط . وكذلك تضخم الكبد. أظهر التحاليل الكيميائي الحيوي زيادة فى قيم إنزيمات الكبد (الفوسفاتيز القلوي، الألانين أمينو ترانسفيريزو أسبارتات أمينو ترانسفيراز) بالإضافة إلى اليوريا، الكرياتينين، حمض البوليك، البروتين، الألبومين، الجلوبيولين، الجلوكوز، الكورتيزول، وإنزيم نازع لهيدروجين اللاكتات بالدم و مستويات الدهون في المجموعة الثالثة المعرضة ل ٥٠ مجم/ لتر من جزيئات الفضة النانوميترية لمدة ٣٠ يومًا .

سجلت معاملات الإجهاد التأكسدي في أنسجة الكبد والكلي ما يلي:

أدى التعرض ل ٥٠ مجم/ لتر من جزيئات الفضة النانوميترية في سمك القرموط الأفريقي إلى إرتفاع في الأكسدة الفوقية للدهون في أنسجة الكبد والكلي. فى حين أن مستوى الجلوتاثيون ،إنزيم الكتاليز، سوبرأكسيد ديزميوتيز ، القدرة الكلية المضادة للاكسدة قد زاد مقارنة بالأسماك الضابطة بعد التعرض ل ٥٠ مجم/ لتر من جزيئات الفضة النانوميترية لمدة ٣٠ يومًا.

نتجت تعرض جزيئات الفضة النانوميترية فى زياده تلف الحمض النووي للمجموعة الثالثة المعرضة لجزيئات الفضة النانوميترية بالمقارنة مع مجموعة التحكم والعلاج بحبة البركة.

تضمنت التغييرات النسيجية المرضية الأكثر ملاحظة ما يلي:

- **الكبد :** تراكم خلايا الكبد مع إلتهاب خلوي بالمسالك البابيه ووجود تليف وإلتهاب بخلايا الكبد وبؤر متعدده
- **الكلي :** وجد تغيرات وإلتهابات بمحافظ بومان وتغيرات فى خلايا الأنابيب الكلوية وقد ظهر إرتشاح خلوى التهابي.

الخلاصة:

حسنت التغذية بحبة البركة لسمك القرموط الأفريقي من الأثار الجانبية لسمية الفضة النانوميترية وتحسين المتغيرات السابقة بالقرب من المجموعة الضابطة. كشفت الدراسة الحالية أن النظام الغذائي بحبة البركة له تأثير جيد ضد سمية الفضة النانوميترية في سمك القرموط الأفريقي.

المستخلص العربى

تأثير حبة البركة على جزيئات الفضة النانوميترية المحدثة للتسمم الكبدي والكلوي فى أسماك القرموط

تستخدم جزيئات الفضة النانوميترية بشكل فعال في تطبيقات تربية الأحياء المائية المتنوعة. هدفت هذه الدراسة الى تعيين التركيز النصف مميت في أسماك القرموط الأفريقي بالإضافة إلى الدور الوقائي لحبة البركة ضد السمية المحدثة من جزيئات الفضة النانوميترية.

تم تقسيم الأسماك إلى ٤ مجموعات كل مجموعه تكررت ثلاث مرات كالأتى: المجموعة الأولي وهى المجموعة الظابطة ،المجموعة الثانية تم تغذيتها على ٣% حبة البركة، المجموعة الثالثة تم تعرضها إلى ٥٠ ملجم/ لتر من جزيئات الفضة النانوميترية ، المجموعة الرابعة تم تعرضها إلى ٥٠ ملجم/ لتر من جزيئات الفضة النانوميترية وتغذيتها على ٣% حبة البركة لمدة ٣٠ يوماً. في نهاية التجربة تم جمع عينات الدم وأنسجة الكبد والكلي من الأسماك للتحليلات البيوكيميائية و النسيجيه.

أظهرت النتائج زياده ملحوظة في إنزيمات الكبد (الفوسفاتيز القلوي، الألانين أمينو ترانسفيريزو أسبارتات أمينو ترانسفيراز) وأنشطه الكلي (اليوريا، الكرياتينين ،و حمض البوليك). سجلت النتائج إرتفاع في مستويات البروتينات الكلية، الألبومين، الجلوبيولين، ونسبة الألبومين / الجلوبيولين. كذلك في مستويات الجلوكوز و الكورتيزول ،و إنزيم نازع لهيدروجين اللاكتات بالدم. وتم تسجيل مستويات الدهون الكلية (الكوليسترول والدهون الثلاثيه).

من ناحية أخري أظهرت التحاليل أن أسماك القرموط الأفريقي الذى تعرضت لجزيئات الفضة النانوميترية قد أحدثت إرتفاع كبير فى الاكسدة الفوقية للدهون فى أنسجه الكبد والكلي.

سجل هذا الإرتفاع أيضا أن مستوى الجلوتاثيون ; انزيم الكتاليز ; سوبر اكسيد ديزميوتيز ;القدرة الكلية المضادة للاكسدة مقارنة بالأسماك الضابطة بعد التعرض ل ٥٠ ملجم / لتر من جزيئات الفضة النانوميترية لمده ٣٠ يوماً.

علاوه علي ذلك، زادت بشكل ملحوظ نسبة تلف الحمض النووي فى كل من أنسجه الكبد والكلي فى سمك القرموط الأفريقي المعرض ٥٠ ملجم/ لتر من جزيئات الفضة النانوميترية.

وأظهرت الدراسة النسيجية للأنسجة الكبدية والكلوية لأسماك القرموط الأفريقى تغييرات وأضرار ملحوظة نتيجة للتعرض ل ٥٠ ملجم/ لتر من جزيئات الفضة النانوميترية لمده ٣٠ يوماً مقارنة بالمجموعة الضابطة والمجموعة المعالجة ٣% بحبة البركة. استنتج من هذه الدراسة أن تغذية أسماك القرموط الأفريقي على ٣% حبة البركة قد حسنت من التاثيرات المدمرة لسمية جزيئات الفضة وقربها من المجموعة الضابطة.

FSC
www.fsc.org
MIX
Papier | Fördert
gute Waldnutzung
FSC® C083411

Kolibri 360 GmbH
Ferdinand-Jühlke-Straße 7
99095 Erfurt, Deutschland
produktsicherheit@kolibri360.de

Druck:
CPI Druckdienstleistungen GmbH
im Auftrag der
Kolibri 360 GmbH
Ferdinand-Jühlke-Str. 7
99095 Erfurt